Perspectives on School Response

This book offers a unique collection of narrative case studies that capture the responses of mental health professionals to tragedies in schools and are designed to connect key concepts and skills with real-life application. By citing evidence-based theories and interventions with vivid real world accounts, this volume highlights the multi-phased, multi-disciplinary nature of school crisis response while emphasizing the need for effective coordination and collaboration. It provides a powerful professional development resource for school crisis teams, psychologists, counselors, social workers, nurses, resource officers, administrators and teachers, and training university students, who will face similar situations.

Jeffrey C. Roth, PhD, NCSP, retired after over twenty years of co-leading a school district crisis response team. He is a licensed psychologist and adjunct professor at Wilmington University and was awarded School Psychologist of the Year by the Delaware Association of School Psychologists in 1993. He volunteers with American Red Cross Disaster Mental Health.

Benjamin S. Fernandez, MS Ed, serves as a lead school psychologist for Loudoun County Public Schools and chair of the National Association of School Psychologists' (NASP) School Safety and Crisis Response Committee. He was named School Psychologist of the Year in 2012 by NASP.

Perspectives on School Crisis Response

Reflections from the Field

Edited by
Jeffrey C. Roth and
Benjamin S. Fernandez

Routledge
Taylor & Francis Group

NEW YORK AND LONDON

First published 2018
by Routledge
711 Third Avenue, New York, NY 10017

and by Routledge
2 Park Square, Milton Park, Abingdon, Oxon, OX14 4RN

Routledge is an imprint of the Taylor & Francis Group, an informa business

© 2018 Taylor & Francis

Library of Congress Cataloging-in-Publication Data
Names: Roth, Jeffrey C., editor.
Title: Perspectives on school crisis response : reflections from the field /
edited by Jeffrey C. Roth and Benjamin S. Fernandez.
Description: New York, NY : Routledge, 2018. | Includes index.
Identifiers: LCCN 2017055444 | ISBN 9781138236905 (hardcover : alk. paper) |
ISBN 9781138236912 (pbk. : alk. paper) | ISBN 9781315301471 (e-book)
Subjects: LCSH: School crisis management. | Crisis intervention
(Mental health services)
Classification: LCC LB2866.5 .P47 2018 | DDC 363.11/9371—dc23
LC record available at https://lccn.loc.gov/2017055444

ISBN: 978-1-138-23690-5 (hbk)
ISBN: 978-1-138-23691-2 (pbk)
ISBN: 978-1-315-30147-1 (ebk)

Typeset in Goudy
by Florence Production Ltd, Stoodleigh, Devon, UK

Contents

About the Editors

Jeffrey C. Roth, PhD, NCSP, joined the Brandywine School District in Delaware where he initiated Operation Cooperation—a school-wide approach to conflict resolution, helped initiate and lead the district Crisis Response Team, and was named School Psychologist of the Year. He developed Instructional Support and Positive Behavior Support Teams, a social skills curriculum, and a program of self-directed professional development for school psychologists. He was trained in the PREPaRE Model, Critical Incident Stress Management, Crisis Prevention Institute (CPI), and Red Cross Disaster Mental Health. He is an American Red Cross Disaster Mental Health volunteer. Dr. Roth earned his doctorate at Temple University. During graduate school, he worked as a recreational therapist at St. Christopher's Hospital for Children and taught graduate courses at Arcadia University. He is a licensed psychologist and adjunct professor at Wilmington University, Delaware. He co-developed a mentoring program for new school psychologists for the state Department of Education. Dr. Roth authored *School Crisis Response: Reflections of a Team Leader* (2015).

Benjamin S. Fernandez, MS Ed, completed his graduate studies in School Psychology at Bucknell University and has practiced as a school psychologist in Pennsylvania and Virginia for 20 years. He serves Loudoun County Public Schools as a lead school psychologist providing a range of diagnostic, prevention, and psychological services as well as coordinating the school psychology internship program. He is a PREPaRE trainer responsible for training school psychologists, interns, and school social workers. He is a crisis team leader and coordinates crisis teams and intervention services. Mr. Fernandez has contributed to presentations, articles, and books related to youth suicide, PREPaRE, and school safety and crisis response. He conducts workshops on PREPaRE and on school safety and crisis response on local, state, and national levels. He has testified before the Senate on a panel for *Examining Mental Health: Treatment Options and Trends*. He was named School Psychologist of the Year by the Virginia Academy of School

Psychologists and in 2012, was named School Psychologist of the Year by the National Association of School Psychologists (NASP). In 2015, Mr. Fernandez received a NASP Presidential Award. He is chair of NASP's School Safety and Crisis Response Committee.

About the Contributors

Kathy Bobby, MA, EdS, NCSP, served as a school psychologist for the Fort Mill and the Charleston County school districts in South Carolina. Ms. Bobby is a Nationally Certified School Psychologist, a Licensed Psycho-Educational Specialist, and a certified PREP<u>a</u>RE trainer. She has published and presented locally, regionally, and nationally on crisis response and intervention.

Sara Castro-Olivo, PhD, NCSP, is an Associate Professor in the School Psychology Program at Texas A&M University. Her research focuses on the development and validation of culturally responsive social-emotional and behavioral interventions for Latino ELL youth and families, and she has authored numerous chapters and research articles. Dr. Castro-Olivo has extensive experience as an interventionist for Spanish-speaking populations in Texas, California and Oregon, working as a bilingual community health educator, social skills trainer, parent trainer, school psychologist, and mental health care provider.

Christina Conolly, Psy.D, NCSP, is the Director for Psychological Services with the Montgomery County Public Schools in Rockville, MD. She is the former Director for Crisis Intervention and Safety for another school district, where she supervised the suicide and bullying prevention programs, the mental health response team, crisis plan development, staff training, and consulted with schools and community agencies. Dr. Conolly is a member of the National School Safety & Crisis Response Committee, co-author of the *PREP<u>a</u>RE Workshop 1: Crisis Prevention and Preparedness* (2nd ed.), co-author of *School Crisis Prevention and Intervention: The PREP<u>a</u>RE Model* (2016), and articles and chapters on school crisis response.

David J. Denino, M.S., is emeritus faculty from Southern Connecticut State University, where he directed the counseling center, and was part of the Red Cross Disaster Mental Health team that traveled to communities near New Orleans after Hurricane Katrina. He has deployed to other disasters including Super Storm Sandy and the school shootings at Sandy Hook

Elementary. Mr. Denino is currently the lead for Red Cross Disaster Mental Health in Connecticut and Rhode Island.

Cynthia Dickinson, NCSP, serves as program manager, coordinator of crisis response, and dropout prevention for Fairfax County Public Schools. She is a PREPaRE curriculum trainer of over 175 district employees and has addressed crisis incidents as diverse as the events of 9/11, mass shooting at Virginia Tech, natural disasters, and more typical school system responses to homicides, suicides, accidents, and other tragic events. She has served NASP as co-chair of the National Emergency Assistance Team, chair of the School Safety and Crisis Response Committee, represents NASP in the American Red Cross national partners group and is an active disaster responder, and was presented the NASP Award of Excellence in School Safety and Crisis Response.

J. Douglas DiRaddo, Ed.D., C.S.P., serves as a member of the National Association of School Psychologists' National Emergency Assistance Team (NEAT),—responding to crisis situations nationally and internationally. He holds certifications and is a master level trainer for the Crisis Prevention Institute (CPI) in both Nonviolent Crisis Intervention and Autism Spectrum Disorder CPI, and trainer for the PREPaRE curriculum. Dr. DiRaddo is co-leader of the Brandywine School District Crisis Response Team in Wilmington, Delaware and was awarded School Psychologist of the Year by the Delaware Association of School Psychologists.

Terri A. Erbacher, Ph.D., is a school psychologist with the Delaware County I.U. in Pennsylvania, and Clinical Associate Professor at the Philadelphia College of Osteopathic Medicine. She is past president of the local chapter of the American Foundation for Suicide Prevention, serves on the executive committee of the state Youth Suicide Prevention Initiative and chairs the Crisis Prevention Committee of the state Association of School Psychologists. Dr. Erbacher is an author, speaker, trainer, and consultant after critical incidents, specializing in school crisis management, suicide prevention and risk assessment, and grief and traumatic loss. She is lead author of *Suicide in Schools* (2015) and has received service awards from the Delaware County Suicide Prevention Task Force, Survivors of Suicide, American Foundation for Suicide Prevention, and was named Pennsylvania's School Psychologist of the Year.

Melissa Allen Heath, Ph.D., NCSP, Associate Professor in Brigham Young University's School Psychology Program, researches school-based crisis intervention, particularly focusing on the support provided by teachers, staff, and parents. Other research topics include children's grief, bullying, and bibliotherapy. Dr. Heath is a licensed psychologist (UT), certified school psychologist (UT), and Nationally Certified School Psychologist.

Lea Howell, MS, is a school psychologist in the Glendale Unified School District in California. Upon earning her teaching credential in education at Vanderbilt University, she moved to California and started teaching kindergarten. Intrigued by the varying needs of her students, she pursued graduate work in the field of school psychology and now enjoys working with adolescents at a large high school.

Cathy Kennedy-Paine, NCSP, is a retired school psychologist in Springfield, Oregon, and the current Lead for the National Association of School Psychologists' National Emergency Assistance Team (NEAT). She responds to crises in schools across the country and serves as a consultant to communities experiencing school violence. Ms. Kennedy-Paine is an adjunct faculty member at the University of Oregon, has served on the advisory boards for the Oregon Center for School Safety and the National Resource Center for Safe Schools, was an invited panelist at a White House School Safety Summit, and has published numerous articles on school safety and crisis response.

Richard Lieberman, MA, NCSP, is a school psychologist and lecturer in the Graduate School of Education at Loyola Marymount University. From 1986–2011 he coordinated Suicide Prevention Services for Los Angeles Unified School District. Mr. Lieberman is a founding member of the National Association of School Psychologists' (NASP) National Emergency Assistance Team (NEAT), and co-author of *School Crisis Prevention and Intervention: the PREPaRE Model* (2009) and *Best Practices in Suicide Intervention* (2014, 2008 & 2002).

Gabriel I. Lomas, Ph.D., is an Associate Professor of counseling at Western Connecticut State University. Trained in both the NOVA and PREPaRE models of crisis response, he has responded to crisis events in schools and organizations in Texas, New York, and Connecticut. In 2013, he established the Western Connecticut Regional Crisis Team, a collaborative of helping professionals from 13 school districts.

Anthony Pantaleno, Ph.D., NCSP, is a retired School Psychologist whose professional interests include the development of peer helping models, assessment and treatment of personality disorders, application of mindfulness and Buddhist psychology practices, and the preservation of the field of School Psychology. He coordinates activities of the Long Island School Practitioner Action Network (LISPAN) related to school crisis prevention and intervention, and was the NASP 2013 School Psychologist of the Year and the 2008 NYASP Practitioner of the Year.

Dr. Scott Poland, NCSP, Professor, Co-Director of the Suicide and Violence Prevention Office at Nova Southeastern University, is an internationally recognized expert on school crisis and youth suicide. He has authored five books on these subjects, including the recently co-authored *Suicide in Schools* (2015). Dr. Poland has assisted many schools after tragedies including school shootings, suicides, terrorism and natural disasters, and was first chairperson of the National Emergency Assistance Team, past Prevention Division Director of the American Association of Suicidology, and past President of the National Association of School Psychologists.

Melissa A. Reeves, Ph.D., NCSP, LPC, school psychologist, licensed special education teacher and counselor. Dr. Reeves is an Associate Professor at Winthrop University, S. C. She trains nationally and internationally in crisis prevention and intervention, threat and suicide assessment, the impact of trauma and PTSD on academic achievement, and establishing a positive behavior support and response to intervention model. Dr. Reeves is a member of the National School Safety & Crisis Response Committee, lead co-author of the *PREPaRE Workshop 1: Crisis Prevention and Preparedness* (2nd ed.), co-author of *School Crisis Prevention and Intervention: The PREPaRE Model* (2016), and has written numerous articles, chapters, and books on school crisis response. Dr. Reeves served as President of the National Association of School Psychologists 2016–2017.

Jeffrey C. Roth, Ph.D., NCSP, helped initiate and lead his district Crisis Response Team in Wilmington, Delaware, and was named state School Psychologist of the Year. He was trained in the PREPaRE model and Red Cross Disaster Mental Health. He is a Red Cross Disaster Mental Health volunteer, licensed psychologist, and adjunct professor at Wilmington University, Delaware.

Andrew Vengrove, Ed.D., is a School Psychologist in the Westfield Public Schools, Massachusetts where he helped to initiate policies and procedures for crisis response, and serves on the district-wide and middle school crisis teams. Dr. Vengrove is Core Faculty at Union Institute & University where he teaches in their graduate Clinical Mental Health Counseling Program and is a Visiting Professor at Westfield State University. He also holds School Adjustment Counselor/School Social Worker and Approved Clinical Supervisor (ACS) certifications, and is a Licensed Educational Psychologist.

Foreword

Crisis events present both danger and opportunity. There is no doubt that crises can have devastating consequences for individuals, schools, and communities. Responding to them is an intense, emotionally draining experience fraught with unanticipated events, yet efforts of a well-trained, multidisciplinary crisis response team can support resilience, minimize traumatic impact, and help schools return to a new normal.

This book epitomizes the dangers and opportunities through honest and thoughtful narratives from professionals with first-hand experience in school crisis response. Introductions to each section provide guidance, based on theory and research, about the unique facets of each type of crisis and useful application of best practices in prevention, preparedness, response, and recovery. A wide range of crisis events are addressed, including accidents, death of staff and students, suicide, natural disasters, violent death, and child maltreatment. Each section includes personal narratives with detailed depictions of response to crises, in the powerful and honest voices of the caregivers who experienced them. Readers will glean important knowledge through these first-hand accounts and the very helpful lessons learned and reflections provided by authors at the end of each chapter.

Jeff Roth and Ben Fernandez have developed a resource that covers complex topics of school crisis response in an incredibly authentic way. This collection confronts some of the most challenging issues in crisis response, including memorials, complicated grief, funeral attendance and support, the role of social media, coping effectively with national media, contacting families of victims, supporting teachers, reducing contagion, and legal considerations. Authors often provide specific resources helpful in their responses, including actual examples of letters to the school community and outlines of content shared in meetings with staff and students. Many of the authors illustrate and reference how the PREPaRE model (Brock et al., 2016) can be applied specifically to unique crisis events. Experienced, courageous, compelling voices weave important themes of resilience and hope, culturally competent crisis intervention, and caring for the caregiver throughout the book. This collection is essential reading for university students preparing to provide school-based mental health, and members of school crisis response teams.

Amanda B. Nickerson, Ph.D., NCSP, Professor and Director, Alberti Center for Bullying Abuse Prevention, University of Buffalo, Co-Author, NASP PREPaRE School Crisis Prevention & Intervention Curriculum.

Acknowledgments and Dedications

The editors wish to acknowledge the encouragement of this project's development by Dr. Melissa Reeves, Dr. Jacqueline Brown, Dr. Terri Erbacher, and Dr. Melissa Heath. We wish to thank our valued colleagues, who contributed powerful, passionate, informative narratives, without which there would not have been a book to edit. We appreciate the kind support of our Routledge editors, Amanda Devine and Chris Teja, and editorial assistants, Nina Guttapalle and Emma Starr.

This book is dedicated to the memory of Dr. Patricia Minuchin, beloved professor, author, and mentor, whose dedicated work improved the lives of students, educators, and families, and to the memory of Central High School friend Harold Rosenthal, killed by terrorists in an airport while on a diplomatic peace mission.

I have been inspired by Carol and our children and loved ones Leah, Raleigh, Michael, Esther, Maritsa and Aaron.

Jeffrey C. Roth

I have been honored to work with many crisis teams and team leaders and I would like to dedicate this book to the crisis teams and team leaders of Loudoun County Public Schools. I am inspired by their selfless dedication and expertise to the care of students and staff.

To my wife, Michelle, who is my primary support network and my daughters, Meredith and Jill who remind me everyday of why I do what I do.

Benjamin S. Fernandez

Introduction
Perspectives on School Crisis Response

This is a book of stories. Stories of sorrow. Stories of hope. Stories of people helping people. In the tradition of Native American and other tribal cultures, this book uses storytelling as the medium for learning from what happened in the past, and teaching about what may happen again.

THE POWER OF THE PERSONAL NARRATIVE

The ancient Greeks recognized the connection between writing and emotional benefit, naming Apollo the god of both poetry and healing. The English playwright, Shakespeare, demonstrates an uncanny understanding of despair in his tragedy, *Macbeth*, when he writes, "Give sorrow words. The grief that does not speak, whispers of the o'er-fraught heart, and bids it break." Centuries after the bard, psychologists recognized the healing power of the "talking cure"—telling the story of emotional trauma as a means toward recovery. Creative expression, whether through the spoken or written word, art or dance, can support the quest for meaning and healing.

Therapeutic Talking after Traumatic Events

In the late nineteenth century, Sigmund Freud (1901, 1917) developed a therapeutic approach in which patients talked about their deepest thoughts and feelings, reversing adverse effects of repressed traumatic memories. Therapists influenced by Rogers (1951) believe that the process of describing problems with an empathic listener allows clients to arrive at insights that can normalize feelings associated with a traumatic incident, and encourage the client's perception of positive personal qualities, actions, and coping.

Mitchell (1983) recognized the occupational hazard of first responders like police, firefighters and emergency medical services, who are at risk for *vicarious trauma*, or *compassion fatigue*—stress reactions of those who often witness severe trauma as part of caring for victims of terrible events (Figley, 2002). Mitchell recognized the need for approaches to ameliorate these damaging effects and plan ways to manage stress.

Narrative Therapy utilizes clients' stories to understand their perceptions and address problems (Neimeyer, 2009; White, 2007). Stephen Madigan (2011) discusses children's reactions and vulnerability to trauma. The child or adult must never be shamed or forced to tell details of trauma, and there should be sensitivity to cultural norms of expression. Given these cautions, children and adults can discover safe ways to articulate the experience of trauma and find renewed hope.

Written Personal Narratives about Traumatic Events

Viktor Frankl's book, *Man's Search for Meaning* (1997), is a remarkable personal narrative that describes his survival in Nazi concentration camps during World War II. James W. Pennebaker (1997, 2004) asserts that the writing paradigm is an exceptionally powerful therapeutic means of expressing emotional experiences, leading to improved health and a more positive outlook. Writing about trauma can be a tool to assimilate events and discover meaning and self-understanding.

Louise DeSalvo (1999) cautions that while writing about traumatic events can be a "helpful way of integrating them into our lives, of helping us feel happier, of improving our psychic and physical well-being," writing about extreme trauma that has caused deep psychological pain poses the risk of re-traumatization. Writers who have histories of extreme trauma or appear emotionally vulnerable should have skilled therapeutic support.

Narratives as an Educational Tool

Narratives written by crisis responders describing traumatic incidents can be therapeutic for the writers and instructive for prospective responders. The educational aspect of narratives is most effective when individuals or teams reflect upon and internalize the event, creating an orientation that supports resilience and the ability to function more calmly during chaotic circumstances. While evidence-based frameworks such as PREPaRE (Brock et al., 2016) contribute concepts and intervention skills, narratives provide guidance on how to apply concepts and skills to real crises.

Personal narratives that illustrate the problems presented by traumatic incidents in schools can educate and strengthen both new and experienced crisis responders. Compassionate stories in the voice of responders combine an intellectual understanding of how to intervene with an emotional understanding of how it feels to do it. They provide a powerful training tool for the preparation of practitioners and teams who confront the challenges and problem-solving demanded for effective intervention.

Study groups and school or district crisis teams can adapt narratives to conduct "table top" drills. Narratives provide school-based mental health providers and educators with information that helps caregivers prepare for crisis-generated problems. In her book of narratives about catastrophic school crises, Mears (2012) states that the purpose is "to help you prepare for the worst that you can imagine, so if it occurs, you will find yourself able to make the kinds of decisions that need to be made."

Narratives can help prepare for what to expect, what to do, and how to care for self and other caregivers during and after a traumatic event. Bolton and colleagues (2004) suggest practitioners use "writing processes to enable them to reflect effectively upon their practice for professional development."

PREP_a_RE: A MODEL FOR SCHOOL CRISIS RESPONSE

Many of the personal narratives within this book apply concepts and skills from the comprehensive PREPaRE model (Brock et al., 2016). Some narratives about events that happened before, or during the early years of PREPaRE development, apply its concepts and lessons for educational purposes. The PREPaRE curriculum presents a range of coordinated, evidence-based crisis planning, prevention, intervention, and recovery.

PREPaRE emphasizes that trained school safety and crisis teams, and school-based mental health providers must be involved in the following hierarchical and sequential set of activities:

P—**Prevent** and prepare for psychological trauma.
R—**Reaffirm** physical health and perceptions of security and safety.
E—**Evaluate** psychological trauma risk.
P—**Provide** interventions.
a—and
R—**Respond** to psychological needs.
E—**Examine** the effectiveness of crisis prevention and intervention.

ESSENTIAL ELEMENTS OF SCHOOL CRISIS RESPONSE

School crisis prevention, preparedness, and intervention require special knowledge and skills, given the vulnerability of young people, the prevalence of traumatic incidents, the nature of relationships in schools, and the unique consequences of school crises. Rossen and Cowan (2013) emphasize that schools can provide trained, caring adults, an environment that develops resilience, and partnerships with families and community resources that offer systems of support (Blaustein, 2013; Nickerson & Heath, 2008; Reeves et al., 2010; Reeves et al., 2012).

This section presents an overview of essential elements of school crisis response described in the personal narratives of this book.

Crisis Prevention, Preparedness, and Response

School Climate

A positive school climate promotes interpersonal competence and academic success. School-wide programs such as Positive Behavior Support (PBS), anti-bullying, suicide prevention, and social-emotional learning (SEL) can help establish norms that value diversity, shared influence, creative expression, prosocial relationships, and conflict resolution. Students can make academic and extra-curricular choices, set cooperative goals, and experience inclusion and connectedness (Bear, 2010; Cowan et al., 2013; Doll et al., 2014; Sugai & Horner, 2009).

Schools can support active student participation in safety planning and taking responsibility for their school. When schools encourage a climate where students feel a sense of belonging and security, they are better able to learn. Promoting connectedness and resilience produces individuals who are less vulnerable to crisis reactions and better able to cope, recover, and return to learning (Brock et al., 2016; Doll et al., 2014; Joyce & Early, 2014; Masten, 2014; Osher et al., 2012).

Teamwork and Training to Provide Services

School administrators and mental health providers can team to de-stigmatize seeking mental health services, and provide teachers, students, staff, and families with training and consultation to recognize and address emotional trauma, grief, stress reactions and mental illness (Eagle et al., 2015; Kennedy-Paine et al., 2014; Rossen & Cowan, 2013).

Trained school and district crisis response teams are also capable of serving as school safety and threat assessment teams, joining with other educators and teams to identify security needs and take preventive measures. A district level response team is necessary when a school team is overwhelmed by a traumatic incident. Connections with community, regional, and national resources are also valuable (Reeves et al., 2010). System-wide programs of training and consultation can prepare for response to traumatized and bereaved students, staff, and community (Brown & Jimerson, 2017; Poland et al., 2014).

A Framework for Strengthening Response Capacity

Entering tragic situations with a conceptual framework is crucial, especially when circumstances require that plans be adapted or changed. Having a practiced structure allows smoother adjustments when dealing with the unexpected while under duress. Flexibility, creativity, and "thinking on the spot" are all enabling qualities (Roth, 2015). PREPaRE describes an Incident Command System (ICS) with a shared leadership structure and functional roles to address the varied demands of school crises. Crises present opportunities to review and strengthen safety plans and response capacity.

Developmental, Cultural, and Special Needs Competence

Stage of development influences children's understanding of trauma, reactions, coping style, and memory of the incident. Severe or repeated trauma can impact development and increase children's future vulnerability (Jimerson et al., 2012; Zenere, 2009).

Culturally competent planning for crises includes a cultural needs assessment. Consulting with community leaders and having responders who are familiar with the ethnic, racial, and religious groups in the school, are proficient in languages of the community or have access to interpreters are critical needs. Educators and responders can be trained to be sensitive to diverse expressions of need (Ortiz & Voutsinas, 2012).

Students with special needs can be especially vulnerable and require thoughtful planning that anticipates ways they might react to a traumatic incident. Preparation can be based on strategies that have worked in the past to reassure safety and promote constructive expression of emotions (Clarke et al., 2014; Susan, 2010).

Media Plans and Cooperation

A plan for setting boundaries to prevent media intrusion on campus is crucial, but cooperation with the media to minimize trauma exposure and provide useful information to promote coping is mutually beneficial.

A media plan can include:

1) Periodic briefing for media at an off campus location.
2) Spokespersons skilled in communication that excludes sensationalism.
3) Written statements describing actions to address needs and problem-solve to reduce present and future emotional trauma.
4) Strategies and caregiver training to shield students and help manage print and visual media to minimize trauma exposure.

Systems Thinking

Think systemically throughout the response, keeping families, educators, and other schools informed. Encourage, involve, and empower students and staff when there are opportunities for constructive action. School counselors, psychologists, administrators, social workers, nurses, teachers, and families can intervene on many levels to help students replace despair with hope, creativity, and cooperation (Roth, 2015).

Elements of Response to a School Crisis

Mobilization: Initial Phase of Response

The severity of a traumatic incident determines the level of response necessary to provide sufficient resources for the affected school community.

Severity of incident determining variables:

1) *Predictability*—sudden onset is generally most severe.
2) *Intensity*—large-scale event on campus with witnesses is most severe.
3) *Consequences*—resulting injuries, deaths, destruction of property.
4) *Duration*—time length of trauma exposure, including media images.

Level of Response—Determining the correct level of response is critical since students get cues affecting their perception of severity from the extent of response and the demeanor of adults. Over-response or under-response can cause harm. More responders than necessary can increase perceived trauma, while fewer than needed provides insufficient resources to support a distressed community. Flexibility allows district level responders to be dismissed or "on-call" depending on need.

Levels of response include:

1) *Minimal*—Normal roles of school staff is sufficient to meet needs.
2) *Building level*—School crisis team without outside help is sufficient.
3) *District level*—Blend of school and district teams is sufficient.
4) *Regional/Community level*—Catastrophic events may require support from agencies beyond the school district.

Source: Adapted from Brock et al., 2016
and DiRaddo & Brock, 2012

Crisis Team Planning: Initial Briefing

Initial briefing usually happens early in the morning, before students arrive and prior to a faculty meeting. Depending on the time of the incident, the briefing may be at the end of the school day, over the weekend, or as soon as possible after psychological first aid when there is a sudden incident.

Initial briefing plans generally involve:

1) Gathering and sharing information to separate rumors from facts.
2) Discussion of what to share with stakeholders—teachers, staff, and families. Teachers want to be informed so they can care for their students.

3) Discussion of what to say and not say to students and how to inform them—often a script for classroom teachers with responder support if needed. A script is helpful for office staff response to phone inquiries. PA announcements and assemblies are not recommended since follow-up with distressed students can be extremely difficult.

4) Determine location for interventions—school libraries can be structured for group interventions. Libraries usually have adjoining rooms for *homogeneous* small group and individual crisis counseling.

5) Initiate **primary triage**—As the response begins, school administrators and counselors can generally identify the most severely affected students needing immediate attention. S*econdary triage* identifies students and staff needing attention throughout the response. Parent training helps to identify and support those needing attention (Reeves et al., 2010). Consider following the daily classroom schedule of a student or teacher who has died or is seriously ill or injured to provide a group intervention. A responder should be designated to contact other schools that might have affected students. Since those most affected are not always obvious, triage also involves a process of *evaluating risk for psychological trauma*. The purpose is to identify those at risk and match the appropriate intervention with their degree of need (Brock et al., 2016).

Variables for evaluating risk include:

- Nature and consequences of the traumatic event.
    - Predictability, intensity, consequences of the event—Was the event sudden or expected? Were there fatalities?
    - Duration of exposure—Was the event lengthy or repeated?

- Personal risk factors.
    - *Physical proximity*—Close to incident? At school? At home? Witnesses?
    - *Emotional proximity*—Close relationship? Family? Friends? Teammates?
    - *Internal personal vulnerability*—Poor emotional regulation? History of trauma, loss or mental illness?
    - *External personal vulnerability*—Family and/or social support systems?
    - *Perception of threat*—Continuing perceived risk?
    - *Early and enduring warning signs*—Maladaptive coping? Symptoms of PTSD?

School Staff Meeting

It is imperative to *stabilize* staff so they can care for their students.

Planning items for the staff meeting:

1) Information is shared about the traumatic incident.
2) What to say to students, when, and how—a script for teachers.
3) What to look for—student reactions indicating need for referral.
4) How and where to refer—escort system to accompany those referred.
5) Handouts for teachers, including discussion topics and activities.
6) Reestablish normal routines while maintaining support—honest *reassurance of safety and security* is paramount.
7) Adjust class schedule and testing, if time is needed to process the loss.
8) *Reestablish support systems*—student, teacher, family, faith. If needed, review plan for reunification with parents or primary caregivers. Consider traffic and crowd management. Provide a comfortable space for *caregiver training*.
9) Offer support to teachers, including *crisis counseling*, classroom support, and substitutes (floating) as needed. Teachers, administrators, and staff can model healthy grieving and expression of feelings for students.
10) Establish tentative time for next staff debriefing, or notify staff by email and/or announcement when it will happen.

Crisis Interventions

Psychological First Aid—Essential aims:

1) Stabilize symptoms—maladaptive reactions are prevented from worsening by:
   - Reaffirming realistic perceptions of safety and security
   - Reestablishing social support systems
   - When there is a severe reaction, grounding techniques such as stress management, deep breathing, orienting to surroundings, or guided imagery for relaxation can help.

2) Reduce symptoms, while normal grieving is supported.
3) Reestablish functional capacity, including adaptive coping, problem solving if ready, or referral for more intensive treatment, if needed.

Psychological First Aid—Guidelines for approaching distressed children:

1) Observe status, then initiate contact without intruding or interrupting.
2) Offer practical assistance (water, food, blankets).
3) Ask simple, respectful questions about how you can help, without making assumptions about their need.
4) Be present and supportive, remaining flexible and understand that not everyone affected will want or need to talk with you.

5) Speak calmly and patiently, without using jargon.
6) Listen carefully to understand what they want to communicate and how to help—children may express needs through behaviors, play, or art.
7) Reinforce personal strengths and coping strategies.
8) Give accurate, age-appropriate information that addresses immediate goals and omits disturbing details.
9) Remember that the goal is to reduce distress, address immediate needs, and promote adaptive functioning—not pressure to talk about trauma.
10) If ready, initiate simple problem solving that addresses stated concerns and enhances strengths, coping, and perceived security.

Source: Adapted from Brymer et al., 2012

Levels of Intervention—for most, recovery is a natural process with existing support systems and minimal or no intervention (Bonnano & Mancini, 2008; Brock et al., 2016). For the typically small percentage more severely affected—often family, friends, witnesses, first responders, and vulnerable populations, monitor their progress and intervene when necessary. Document all student and parent contacts, emotional status, and estimated need for follow-up and referral.

Crisis intervention that is not needed, or does not match the individual's degree of need can cause harm (Brock et al., 2016; Nickerson et al., 2009). In general, small, *homogeneous groups* experiencing similar levels of distress, helps prevent *vicarious trauma* and contagion. Imagine the negative effect of grouping severely traumatized students with minimally exposed students. A range of individual, group, or classroom interventions is matched with students according to the severity of trauma indicated by psychological triage (Brock et al., 2016).

Death by Suicide—postvention requires school and community-wide application of a special set of knowledge and skills to deal with accompanying stigma, complicated grief, and possible contagion. The deceased person can be remembered and the survivors' grief acknowledged without glorifying the act. Mental health providers can train teachers to identify signs of distress in those at risk and refer when needed. It is crucial to de-stigmatize seeking help for depression, substance abuse, suicidal ideation, and other mental health problems, and to provide opportunities to participate in constructive, life-affirming activities. *Suicide in Schools* (Erbacher et al., 2015) is a recommended resource.

Crisis Team Debriefings During Response

The response team should debrief at least once a day. Creativity may be needed to find meeting times, share when and where the meeting will happen, and make a concerted effort to involve administrators in attending the meeting and playing a leadership role in decision-making.

Some purposes of team debriefings:

1) Share information and perceptions of the response process, interventions, and counseling themes to sense needs and plan next interventions.
2) Share names of students seen and conduct *secondary triage*, identifying those appearing most affected during interventions, and determining if follow-up is indicated. Contact parents/guardians of severely affected students to offer information about support, monitoring, and referral if needed. Document all contacts!
3) Brainstorm "to do" list of tasks for responders and administrators:

   – Prepare letters to be sent home.
   – Contact grieving family to express sympathy, determine needs and status of funeral plans, such as public or private.
   – Share information with staff, community, and media.
   – Consider care for team caregivers, network of support, and individual stress management plans.
   – Identify next team debriefing target time and location.

Disengagement: Final Phase of Response

The response team should begin disengagement when the school staff is sufficiently able to care for students progressing toward normal routine. The imperative of disengagement is to leave students in the hands of stabilized, recovering administrators, teachers, and support staff who are capable of caring for them. Teachers should understand that return to classwork does not mean that all students are ready to concentrate fully on academics (Roth, 2015).

The PREPaRE model recommends **tertiary triage** during disengagement, which involves careful consideration of the need for continued monitoring or therapeutic treatment for those severely affected. Continue to support families of affected students, especially after catastrophic incidents such as school shootings or death by suicide. Follow-up may be necessary for months and even years, with special attention to anniversaries (Roth, 2015).

Planning for memorials should be considered prior to disengagement. Reeves and colleagues (2010) discuss guidelines for memorials, reminding us that children's understanding of death depends upon their age and cognitive development.

Suggestions for planning school memorials:

1) Written policies are helpful to ensure equity of observance, but don't assume that "one size fits all".
2) Consider memorials for all student or staff deaths, including death by suicide, substances, or high risk behavior. However, with these deaths,

the act must not be glorified. Use caution when counseling students and memorializing a death by suicide, using best practices to avoid contagion (Erbacher et al., 2015; Hart, 2012; Jellinek & Okali, 2012).

3) A memorial advisory committee of students with adult supervision can ensure that decisions are consistent with policies. A menu with options can support creative expression and empower constructive action.

4) Encourage voluntary participation in developmentally appropriate life–affirming activities such as writing letters and poems, creating a memory book for the grieving family, and supporting mental health organizations.

5) Consistent with *secondary triage*, memorial activities provide an opportunity to identify students at risk or experiencing complicated grief. Activities should be structured so attendance is voluntary.

6) Set time guidelines for short-term memorials and their removal. Timelines for memorial events should not rush grief work or cause re-traumatization. Memorials should avoid recounting trauma or anger, but rather affirm life and foster hope and resilience (Heath et al., 2008; Johnson, 2006).

7) Be culturally sensitive, respecting diversity and religious differences in rituals, traditions and practices (Ungar, 2008).

8) Teachers should be offered counseling and information to care for their students and each other. If a staff member has died, substitutes should provide teachers the opportunity to attend the funeral.

9) Set guidelines for long-term memorials such as tree planting, plaques, and library gifts. Permanent memorials should not be placed in front of the school, but in an accessible location that allows the choice to visit.

10) On the anniversary, consider a voluntary remembrance activity that supports healing. The type and severity of the incident may indicate whether an event should be planned. Staff should be vigilant around anniversary dates for the needs of vulnerable students (Zibulsky, 2012).

Demobilization should include a timely closure debriefing, especially since district team members will be relatively isolated from team support.

Debriefing for Crisis Response Team Closure

Engage in post-response examination of the team's experience to process effectiveness, lessons learned, and facilitate stress management. While each team member is affected differently, all can benefit from peer support and a recovery plan that fits their individual needs (Crepeau-Hobson & Kanan, 2014; Roth, 2015).

Main points of closure debriefing:

1) Response caregivers tell their stories and express their feelings.
2) Discuss difficult aspects of the response, including unique challenges.

3) Discuss what went well, documenting the effectiveness of interventions and data that indicates return to pre-crisis capacity.
4) Discuss what could have gone better, and consider whether policies and procedures need to be modified.
5) Discuss "what did we learn about our team and our response that will be helpful in the future?"
6) How are we taking care of ourselves? Are there individual recovery plans?

REFERENCES

Bear, G.G. (2010). *School discipline and self-discipline: A practical guide to promoting prosocial student behavior*. New York: Guilford Press.

Blaustein, M. (2013). Childhood trauma and a framework for intervention. In E. Rossen and R. Hull (Eds.). *Supporting and educating traumatized students: A guide for school-based professionals* (pp. 3–21). New York: Oxford University Press.

Bolton, G., Howlett, S., Lago, C., & Wright, J.K. (Eds.). (2004). *Writing cures: An introductory handbook of writing in counseling and therapy*. New York: Routledge.

Bonanno, G.A. & Mancini, A.D. (2008). *The human capacity to thrive in the face of potential trauma*. Pediatrics, 121, 369–375.

Brock, S.E., Nickerson, A.B., Louvar Reeves, M.A., Conolly, C.A., Jimerson, S.R., Persce, R.C., Lazzaro, B.R. (2016). *School crisis prevention and intervention: The PREPaRE model* (2nd ed.). Bethesda, MD: National Association of School Psychologists.

Brown, J.A. & Jimerson, S.R. (Eds.). (2017). *Supporting bereaved students at school*. New York: Oxford University Press.

Brymer, M., Taylor, M., Escudero, P., Jacobs, A., Kronenberg, M., Macy, R., Mock, L., Payne, L., Pynoos, R., & Vogel, J. (2012). *Psychological first aid for schools: Field operations guide* (2nd ed.). Los Angeles, CA: National Child Traumatic Stress Network.

Clarke, L.S., Jones, R.E., & Yssel, N. (2014). Supporting students with disabilities during school crises: A teacher's guide. *Teaching Exceptional Children, 46*, 169—178.

Cowan, K.C., Vaillancourt, K., Rossen, E., & Pollitt, K. (2013). *A framework for safe and successful schools* [Brief]. Bethesda, MD: National Association of School Psychologists. Retrieved from: www.nasponline.org/Documents/Research%20and %20Policy/Advocacy%20Resources/Framework_forSafe_and_Successful_School_ Environments.pdf

Crepeau-Hobson, F. & Kanan, L.M. (2014). After the tragedy: Caring for the caregivers. *Phi Delta Kappan, 95*(4), 33–37.

DeSalvo, L. (1999). *Writing as a way of healing: How telling our stories transforms our lives*. Boston: Beacon Press.

DiRaddo, J.D., & Brock, S.E., (2012). Is it a crisis? *Principal Leadership, 12*(9), 12–16.

Doll, B., Brehm, K., & Zucker, S. (2014). *Resilient classrooms: Creating healthy environments for learning* (2nd ed.). New York: Guilford Press.

Eagle, J.W., Dowd-Eagle, S.E., Snyder, A., & Holtzman, E.G. (2015). Implementing a multi-tiered system of support (MTSS): Collaboration between school psychologists and administrators to promote systems-level change. *Journal of Educational and Psychological Consultation, 25*(2–3), 160–177.

Erbacher, T.A., Singer, J.B., & Poland, S. (2015). *Suicide in schools: A practitioner's guide to multilevel prevention, assessment, intervention, and postvention.* New York: Routledge.

Figley, C.R. (2002). *Treating compassion fatigue.* New York: Brunner-Routledge.

Frankl, V. (1997). *Man's search for meaning: An introduction to logotherapy.* Boston: Beacon Press.

Freud, S. (1901). *The psychopathology of everyday life.* (Standard Edition, Vol. 6).

Freud, S. (1917). *Introductory lectures on psycho-analysis.* (Standard Edition, Vols. 15 and 16).

Hart, S.R. (2012). Student suicide: Suicide postvention. In S.E. Brock & S.R. Jimerson (Eds.) *Best practices in school crisis prevention and intervention* (pp. 525–547; 2nd ed.). Bethesda, MD: National Association of School Psychologists.

Heath, M.A., Bingham, R., Dean, B. (2008). The role of memorials in helping children heal. *School Psychology Forum: Research in Practice, 2,* 17–29.

Jellinek, M.S. & Okoli, U.D. (2012). When a student dies: Organizing the school's response. *Child and Adolescent Psychiatric Clinics of North America, 21,* 57–67.

Jimerson, S.R., Stein, R., & Rime, J. (2012). Developmental considerations regarding psychological trauma and grief. In S.E. Brock & S.R. Jimerson (Eds.) *Best practices in school crisis prevention and intervention* (pp. 377–399; 2nd ed.). Bethesda, MD: National Association of School Psychologists.

Johnson, K. (2006). *After the storm: Healing after trauma, tragedy and terror.* Alameda, CA: Hunter House.

Joyce, H.D., & Early, T.J. (2014). The impact of school connectedness and teacher support on depressive symptoms in adolescents: A multilevel analysis. *Children and Youth Services Review, 39,* 101–107.

Kennedy-Paine, C., Reeves, M.A., & Brock, S.E. (2014). How schools heal after a tragedy. *Phi Delta Kappan, 95*(4), 38–43.

Madigan, S. (2011). *Narrative Therapy.* Washington, D.C.: American Psychological Association.

Masten, A.S. (2014). Global perspectives on resilience in children and youth. *Child Development, 85,* 6–20.

Mears, C.L. (Ed.) (2012). *Reclaiming school in the aftermath of trauma: Advice based on experience.* New York: Palgrave MacMillan.

Mitchell, J.T. (1983). When disaster strikes: The critical incident stress debriefing process. *Journal of Emergency Medical Services, 8,* 36–39.

Neimeyer, R.A. (2009). *Constructivist psychotherapy: Distinctive features.* New York: Routledge.

Nickerson, A.B., & Heath, M.A. (2008). Developing and strengthening crisis response teams. *School Psychology Forum, 2*(2), 1–16.

Nickerson, A.B., Reeves, M.A., Brock, S.E., & Jimerson, S.R. (2009). *Identifying, assessing, and treating posttraumatic stress disorder at school.* New York: Springer.

Osher, D., Dwyer, K., Jimerson, S.R., & Brown, J.A. (2012). Developing safe, supportive, and effective schools: Facilitating student success to reduce school violence. In S.R. Jimerson, A.B. Nickerson, M.J. Mayer, & M.J. Furlong (Eds.), *Handbook of school violence and school safety: International research and practice* (pp. 27–44; 2nd ed.). New York: Routledge.

Ortiz, S.O., & Voutsinas, M. (2012). Cultural considerations in crisis intervention. In S.E. Brock & S.R. Jimerson (Eds.) *Best practices in school crisis prevention and intervention* (pp. 337–357; 2nd ed.). Bethesda, MD: National Association of School Psychologists.

Pennebaker, J.W. (1997). *Opening up: The healing power of expressing emotions.* New York: Guilford.

Pennebaker, J.W. (2004). *Writing to heal: A guided journal for recovering from trauma and emotional upheaval.* Oakland, CA: New Harbinger Publications.

Poland, S., Samuel-Barrett, C., & Waguespack, A. (2014). Best practices for responding to death in the school community. In P.L. Harrison & A. Thomas (Eds.) *Best practices in school psychology: Systems-level services.* (pp. 302–320). Bethesda, MD: National Association of School Psychologists.

Reeves, M., Kanan, L., & Plog, A. (2010). *Comprehensive planning for safe learning environments: A school professional's guide to integrating physical and psychological safety— Prevention through recovery.* New York: Routledge.

Reeves, M.A., Conolly-Wilson, C.N., Pesce, R.C., Lazarro, B.R., & Brock, S.E. (2012). Preparing for the comprehensive school crisis response. In S.E. Brock & S. R. Jimerson (Eds.) *Best practices in school crisis prevention and intervention* (pp. 245–264; 2nd ed.). Bethesda, MD: National Association of School Psychologists.

Rogers, C.R. (1951). *Client-centered therapy: Its current practice, implications, and theory.* Boston: Houghton Mifflin.

Rossen, E. & Cowan, K. (2013). The role of schools in supporting traumatized students. *Principal's Research Review, 8*(6), 1–8.

Roth, J.C. (2015). *School crisis response: Reflections of a team leader.* Wilmington, DE: Hickory Run Press.

Shakespeare, W. (1977). *Macbeth.* London: Oxford University Press.

Sugai, G., & Horner, R.H. (2009). Defining and describing schoolwide positive behavior support. In W. Sailor, G. Dunlap, G. Sugai, & R. Horner (Eds.), *Handbook of positive behavior support* (pp. 307–326). New York: Springer.

Susan, M.K. (2010). Crisis prevention, response, and recovery: Helping children with special needs. In A. Canter, L.Z. Paige, & S. Shaw (Eds.), *Helping children at home and school III: Handouts for families and educators* (pp. S9H4–1—S9H4–3). Bethesda, MD: National Association of School Psychologists.

Ungar, M. (2008). Resilience across cultures. *British Journal of Social Work, 38,* 218–235.

White, M. (2007). *Maps of narrative practice.* New York: Norton.

Zenere, F.J. (2009). Violent loss and urban children: Understanding the impact on grieving and development. *NASP Communiqué, 38*(2), 1–9.

Zibulsky, J. (2012). Preparing for the anniversaries of crisis events. In S.E. Brock & S.R. Jimerson (Eds.) *Best practices in school crisis prevention and intervention* (pp. 423–434; 2nd ed.). Bethesda, MD: National Association of School Psychologists.

Section 1

School Crisis Response to Transportation Accidents

INTRODUCTION

What a pleasant surprise when I learned that our daughter Leah, a talented second year high school English teacher, was coming after work to our house for dinner. Traffic was unusually "backed up", but she would soon join Carol and me. After dinner and conversation about her teaching, she prepared to leave our house. Then the cell phone call came. Traffic had been "backed up" because of a horrible accident. Eleventh graders, a teenage boy and girl were killed. The boy was a current student of Leah's; the girl had been her student the previous year. Through her tears she said, "We told him not to speed. We told him to slow down." It was hard to see my daughter experience the kind of pain I had often seen as a school crisis team leader. Leah and I talked. I gave her some literature. She got back on the phone with another colleague and they talked for a long time. I think that helped the most (Roth, 2015).

The suddenness of serious injury or death resulting from transportation accidents can be devastating for a school community. The circumstances of the incident and extent of the impact on students and staff determine the necessary level of crisis response. See book "Introduction", *Elements of Response to a School Crisis*.

School Bus Accidents and Safety

A school bus accident may be defined as any event that results in damage to the bus and/or injury to the driver or passengers. While death is rare, nearly 1000 students per year suffer accident-related passenger injuries. Injuries and fatalities also happen when students are walking to and from a school bus. Pedestrians killed by school buses are often small children on their way home (Srednicki, 2004).

Poor judgment or disruptive behavior by students can distract drivers, who often report being overwhelmed by students' lack of respect and failure to follow rules and regulations. Students, parents, educators, and community members must be aware of school bus related risks and work to prevent accidents. Motorists must know and obey regulations about driving near school buses and drop-off zones. Safety programs include "Operation School Bus Safety" or "Be Cool. Follow the Rules" (Srednicki, 2004).

School Bus Safety Rules and Accident Procedures

On the way to the bus:

- Be alert—arrive at your stop at least five minutes early.
- Always obey all traffic lights and signals.
- Plan to walk with schoolmates whenever possible, facing traffic.
- Always cross streets at crosswalks and intersections.
- Look both ways before crossing the street.

At the bus stop:

- Stand back from the curb.
- Do not push when entering or exiting the bus and always use steps and handrail.
- Always obey bus driver and wait for the driver's signal before crossing street.
- Always cross at least 10 feet in front of the school bus.
- Never, never crawl under the bus.

When riding the bus:

- Take your seat quietly and quickly—remain seated when the bus is moving.
- Keep feet on the floor (if they reach).
- Never extend your hands, arms, head, or any object out the bus window.
- Talk in a quiet voice, be polite to the driver and schoolmates, and avoid misbehavior that might distract the driver.

Source: Adapted from Srednicki, 2004

In the event of a bus accident:

- Fire Marshal or law enforcement is generally in charge at the scene.
- School administrator should designate a school leader and proceed to the scene.
- Record names of students on the bus, where they sat (if possible), and current status, accounting for all students—report list back to school leader and nurse.
- Stay calm and reassure students.

- Depending on severity of the event, and especially if there are fatalities or serious injuries, mobilize the blended school and district crisis response teams.
- If mobilization of crisis team is necessary, proceed with trained procedures for response to a sudden traumatic event.
- District transportation should send a second bus to the scene to bring screened, triaged students deemed okay for return to school.
- Phone to tell school nurse names of any students being transported to hospital(s), and destination if known.
- Accompany injured students to hospital—if more than one hospital, send a staff member who is familiar with students to each hospital.
- Using emergency cards, contact parents/guardians of the injured and give directions to hospital—notify district transportation if a family needs a ride.
- Meet arriving family members, guide to waiting area, and provide information and support as needed.
- Maintain contact with the school and district offices.
- Prepare to evaluate bus passengers returning to school—escort to school nurse to be checked for injury and possible referral to hospital emergency room.
- School counselor, school psychologist, and other responders evaluate emotional status and provide counseling support if needed.
- If information about the accident is spreading by cell phones and causing concern at school, provide a brief report of incident and help being provided at the scene.
- Arrange phone contact with parents/guardians of passengers evaluated at school:

  – provide caller with outline of information to be conveyed, including brief description of the event and status of the individual child;
  – if follow-up medical care is indicated, school nurse should make contact;
  – document time of call, caller's name and person contacted.

- Teachers should be informed at a reasonable time by administrator/designee about the status of their students, and be provided support if student is injured or deceased.
- Suggestions for working with concerned parents/guardians arriving at school:

  – identify a comfortable location to accommodate parents/guardians;
  – establish a sign-out procedure for uninjured students;
  – when concerned parent/guardian comes to school inquiring about child, refer to list of uninjured, injured, or deceased students if available, and if confirmed "uninjured" might say "Your child is not

injured" and direct to pick-up location to be reunited with child and leave school;

– parent/guardian of hospitalized or deceased child should be directed to a private location for information, emotional support, and escort to hospital.

• Consider sending a letter home with affected students summarizing the event, the school's response, and a contact number if there are concerns—consider a follow-up phone call for parents/guardians of injured students unavailable to deliver the letter, documenting all contacts.

• If the event creates the likelihood that passengers will be anxious about riding the bus, assign a supportive staff member to ride on at least the next trip and assess need for further bus or counseling support.

• Depending on event severity, consider a before/after school staff meeting to:

– provide current status of students sent to hospitals, if any;
– acknowledge efforts by teachers, school nurse, and other support staff;
– provide opportunity for questions and processing staff reactions;
– ask staff to identify students or staff in need of emotional support—remind staff about help options, including Employee Assistance Program.

Source: Adapted from *Crisis Response Manual and Guidelines*, Brandywine S. D., DE.

Car Accidents and Prevention

Car accidents can result in death and serious injury to children, adolescents, and sometimes families. The suddenness and unexpected nature of these tragedies intensifies the emotional reaction of the school community.

A lack of mature judgment and driving experience puts teenagers at greater risk for accidents, but that risk is multiplied many times by drinking alcohol, speeding, and/or using drugs. While the percentage of teens in high school who drink and drive has decreased by more than half since 1991, more needs to be done. Nearly one million high school teenagers drank alcohol and drove in 2011. Teenage drivers are three times more likely than more experienced drivers to be in a fatal car crash. Young drivers, aged 16–20, are 17 times more likely to die in a car crash when they have a blood alcohol concentration of .08 or higher. However, research has demonstrated that there are proven steps that can help protect the lives of young drivers and everyone who rides or shares the road with them (CDC, 2012).

What works to prevent teen accidents and specifically, drinking and driving:

• **Parental involvement**—monitoring and restricting what new drivers are allowed to do as they learn to drive. Parents who establish and enforce "rules of the road", which can include creating and signing a parent-teen driving agreement report lower rates of risky driving, traffic violations, and crashes.

- **Graduated driver licensing** (GDL) systems help new drivers gain experience under less risky conditions—teens move through stages, gradually getting privileges such as driving at night or driving with passengers.
- **Increasing teen and parent awareness** of risky behavior through community effort.
- **Minimum legal drinking age** (MLDA) makes it illegal to sell alcohol to anyone under 21 years of age.
- **Zero tolerance laws** make it illegal for those under 21 years of age to drive after drinking any alcohol.
- **Strengthening enforcement** of policies such as minimum legal drinking age, zero tolerance laws, and graduated driver licensing systems.

Source: Adapted from CDC, 2012

School nurses, counselors, pediatricians, and other health professionals can:

- Screen teens for risky behaviors, including:
    - using alcohol, drugs, or other substances;
    - driving after alcohol or drug use;
    - riding with a driver who has been using alcohol or drugs;
    - driving at excessive speeds, or not slowing for poor driving conditions.

- Educate parents and teens about driving risks and especially drinking and driving.
- Encourage parents of new teen drivers to set and enforce "rules of the road" and consider parent-teen driving agreements.
- Remind parents to set an example as safe drivers, before their child is driving.

Source: Adapted from CDC, 2012

Parents of teenage drivers can:

- Understand that most teens who drink want to get drunk.
- Recognize the danger of drinking and driving—that teen drivers are at significantly greater risk of crashing after drinking alcohol.
- Provide teens with a safe way to get home, such as picking them up or paying for a cab, if their driver has been drinking.
- Model safe driving behavior and habits.
- Consider parent-teen agreements to set and enforce "rules of the road".

Source: Adapted from CDC, 2012

Teenage drivers can follow safe driving habits:

- Choose to never drink and drive.
- Refuse to ride with a teen driver who has been drinking or using substances.
- Know and follow state graduated driver licensing laws.
- Follow "rules of the road" in their parent-teen driving agreement.
- Wear a seat belt on every trip, no matter how short.
- Obey speed limits and reduce speed at night and under poor driving conditions.
- Limit nighttime driving.
- Set a limit on the number of teen passengers.
- **Never** use a cell phone or text while driving.

Source: Adapted from CDC, 2012

* * * *

Jay Spilecki was a scholar-athlete, talented, respected and well liked by his high school peers. One tragic night changed everything. When Jay died in a car accident, it was later confirmed that it was related to alcohol consumption. His death became a cautionary tale for young people in the community, largely through the efforts of his parents, who reached out through their grief to cooperate with crisis response and comfort the students. Jay's mother penned a poignant article, reprinted with her permission.

A Short Walk To Deliver A Powerful Lesson

By Sue Spilecki

It's a short walk from the Charter School of Wilmington, where I teach, across Lancaster Pike to the Silver-brook Cemetery. I make that walk every year with about 250 Charter School students—freshmen who are in my Health and Wellness classes. When we leave school, they think they're going on some sort of physical fitness field trip. By the time we return, they've learned a powerful lesson about the dangers of underage drinking and drunk driving— a lesson that I hope stays with them for the rest of their lives.

In 1996, my son, Jay, was a senior in high school. He was a happy, fun-loving kid—not at all different than the kids I see in high school now. That December, he attended a dance with some friends and later ended up at a party where they were drinking alcohol.

I learned that at about 1:45 a.m., he tricked the designated driver into giving him the car keys and he and two friends jumped into a car to check out another party. On Whitby Road in Sharpley, he lost control of the car and hit a tree. He died instantly.

My ninth grade health and wellness students don't know any of this when we set out on our walk across Lancaster Pike. Happy and talkative when we leave the school, they grow quiet when we step on to the grass of the cemetery. By the time I have them gather around my son's grave, they are silent and nervous.

I tell them the lesson is about under-age drinking and driving under the influence of alcohol. I ask if any of them know a family that has been impacted by a drunk driver, and a few always raise their hands. Then I say, "No, you all know someone because you all know me." I then tell them Jay's story. I stress to them that he was an awesome high school student with a bright future and loads of friends and a family that loved him, just like them. And like them, he knew lots of kids who drank and made poor decisions, but nothing bad ever happened, so what was the big deal?

But bad things do happen. I learned that fact in 1996.

Since then, I've learned that two-thirds of Delaware's 11th graders drink alcohol and one in four binge drink. I know that 11th graders who drink are 16 times more likely to get in a car with a driver who is a minor and has been drinking than those who don't drink. I know that they're three times more likely to be forced to have unwanted sexual intercourse, four times more likely to be arrested and six times more likely to use other drugs.

I share these and other statistics with my students. It would be great if they remember these facts, but what I really want them to remember is that they're not indestructible; that bad things can and do happen to kids just like them and that they must not drink and drive.

Parents must remember to be parents to their kids and not buddies. Especially as we move into the "party" season—graduations, beach, "senior week"—all this can translate into tragedy if parents do not step up and actively prevent underage drinking.

Don't serve alcohol to your kids and their friends under the mistaken notion that by keeping them at your house you're keeping them safe. Join me in the effort to enact a "social host" law in Delaware, to hold people responsible for underage drinking on their property.

Parents—and all adults—must play a more active role in preventing underage drinking. You can learn more about teens and drinking by visiting parentsstepup.org, a website established by Delaware's Department of Services for Children, Youth and Their Families.

Or, like my ninth graders, you can learn more by joining me on a short walk across Lancaster Pike.

REFERENCES

Centers for Disease Control and Prevention (CDC). (2012). Vitalsigns. Retrieved from: www.cdc.gov/vitalsigns/teendrinkinganddriving/index.html

English, J., Gordon, L., & Roth, J.C. (1999). *Crisis response manual and guidelines.* Wilmington, DE: Brandywine School District.

Roth, J.C. (2015). *School crisis response: Reflections of a team leader*. Wilmington, DE: Hickory Run Press.

Spilecki, S. (2011). "A short walk to deliver a powerful lesson", *Hockessin Community News*, Delaware. Reprinted with permission.

Srednicki, H.J. (2004). School bus safety: Tips for parents. In A.S. Canter, L.Z. Paige, M.D. Roth, I. Romero & S.A. Carroll (Eds.) *Helping children at home and school II: Handouts for families and educators* (S9–37—S9–38). Bethesda, MD: National Association of School Psychologists.

1 Fatal Car Crash

Tag You're It—Reflections of a New Leader

J. Douglas DiRaddo

INTRODUCTION

Goethe, a German poet and writer, once wrote: "Treat a man as he is and he will remain as he is. Treat a man as he has the potential to become and you make him better than he is." I did not think that applied to me until a small school district in Delaware needed a crisis team to respond to a horrific tragedy. This was a team that had lost its leader to retirement the year before. So, in October 2009, a defining event in my professional career required that I respond to "the potential to become" even though I was content to "remain" as I was. The car crash and death of three out of four family members on that October, 2009 Saturday, required that three other very capable individuals and I, form a core leadership group to respond and lead a well trained team of crisis responders. This was not a straightforward, run of the mill response. They never are. I learned much during the response about professional and personal crisis and how best to respond. It required that the other three members and I form a Core Team that was efficient and effective, and that we do it "in situ", an additional twist that required skills and qualities that the Core Team realized we already possessed. Stephen Covey, in *The Leader in Me* (Covey et al., 2014), makes a simple statement, "Leadership is a choice not a position." During those days of responding as a core team, we found out the choice is not always ours; that sometimes, during the greatest tragedies, you learn the most about who you really are and what you are capable of accomplishing. Through shared leadership working side by side we were able to effect positive change and response through a dark time. Goethe also said, "Knowing is not enough; we must apply. Willing is not enough; we must do." I am glad a small group of administrators pushed the four of us to respond and to do—we are all better for it.

PREPAREDNESS

The Need For a Leader

The summer of 2009 was calm, relaxing, almost too calm. As summer drew to an end and the school year began, it remained calm. Then as homecoming weekend approached and things began to fall into the routine of the school year, it became clear there was a hole in our crisis team. At the end of the prior school year, Dr. Jeff Roth, the leader of the district Crisis Response Team had retired. We had all been trained in PREPaRE and other response models, but functioned solely under the PREPaRE model (Brock et al., 2016). This meant the leadership role was necessary to help mobilize the team, should a crisis occur. Trainer Melissa Reeves made the statement when I was in Anaheim for the initial Pilot project, later known as PREPaRE, that "it is not a matter of IF a crisis is going to happen, but WHEN" (Reeves, 2006). Nowhere was that more true than in our little 11,000 student district. Each year that I served on the team, beginning in 1996, we had mobilized for some crisis response—death of student or staff, fires that resulted in loss of students' lives, suicides, sexual misconduct by a teacher, and homicide of high school students, to name just a few. In each case the team was deployed to the school or schools and the community. Each time, Jeff, a quiet, reserved, very capable leader took the reins and made sure that all aspects of the response were covered. As a result of his leadership, the team moved with confidence and assuredness through each difficult situation. Jeff was very keen on "care for the caregiver" (Creamer & Liddle, 2005; Feinberg et al., 2004) and would continually be checking in with the responders, offering breaks, frequent group debriefings, and one on one check-ins. Jeff had approached me and Leslie Carlson about assuming leadership next year and we agreed. When he retired, and, as the 2009–2010 school year began, the crisis team found itself without its long time leader.

In his book, *School Crisis Response: Reflections of a Team Leader* (2015), Jeff Roth states, "Crisis response is a complex mission requiring multifaceted approaches to leadership. These approaches establish norms that can either enable or obstruct group functioning and effective teamwork." I knew as the year progressed that there was going to be a need for our team to respond, but who would fill that leadership role. Marty Tracy might be a good candidate. He had been working in coordination with Jeff for many years, and was viewed as a leader in the group. However, he didn't view himself as an intervention 'doer', but rather more as a logistics 'getter' (Brock et al., 2016; adapted from the National Incident Management System (US DHS, 2008)). So I began wondering how that role was going to be filled. Other questions began invading my thoughts as well: What would the team structure look like? What if a large number of responders wanted to take a break? Who would be the administrative contact? Would Marty want to also step back? With a new superintendent starting this school year, what would be his vision for the Crisis Response

Team? Would he see its value? Would it even be on his radar, since he was just starting this new role?

The Answer Came Quickly

It was homecoming weekend for several of our district high schools. Saturday, early October, 2009 started out a little hectic. I had just finished cutting the lawn and preparing for the big family "end of summer" party my wife and I had planned. I got in my car to run some last minute errands, and drove to the bakery to pick up the tomato pie I had ordered. Walking through the parking lot my cell phone rang. I answered and Marty Tracy was on the line. He said, "Doug, the superintendent wanted me to call you. We have had a very serious tragedy and we need to get the crisis team together. We want to meet with a few of you today and plan what we should do." We arranged to meet at one of the schools. I called my wife to tell her I was not going to be home. Four of us had been called to meet: Leslie Carlson, Donna Carroll, and Claudette Melton. All three were strong team members, each with their own unique set of skills. Leslie, a coordinator with a school counseling background, brought organization and a level head to the table. Donna and Claudette were both district social workers who worked well with all ages of children and adults. Both were empathetic listeners, skilled in having young children and teenagers open up to them while quickly getting to the heart of the matter. All three were very capable leaders in their own right. And then there was me, a school psychologist in the district.

In March 2006, I participated as the representative from my state organization, the Delaware Association of School Psychologists (DASP) in the pilot version of the PREPaRE curriculum in Anaheim, California. The following year in New York the pilot group met again to finalize the program and become trainers of workshops 1 (Reeves et al., 2011) and 2 (Brock, 2011). Since then I have trained several groups and school districts in Delaware, with fellow trainer Dr. Elliot Davis. Following my PREPaRE training, I took a more active role in training all the members of the district team in that model. The district needed leadership and the four of us were thrust into it in the most profound way. We called ourselves the Core Team, and we would be the ones to fill Jeff's shoes, providing leadership to a 40 member crisis team. Our first response as a Core Team proved to be our most involved, and the need for all 40 members was evident when we heard the news.

DESCRIPTION OF RESPONSE

Why Such an Involved Response?

Saturday afternoon was sunny, but the morning started out foggy and a bit dreary. The four of us sat in a small conference room with Marty Tracy,

Ann Hilkert (Director of Special Education) and Dr. Mark Holodick (the new Superintendent). In fact, because of this tragedy the Board of Education met earlier than scheduled to appoint him Superintendent. Marty started out saying, "We need you guys to pull from all of your experience to help in this situation." He went on:

> This morning the *Prince* family was traveling downstate to go to their son's baseball tournament. In the car were Mr. and Mrs. Prince, *Jason* and his sister *Janice*. As the car was moving through an intersection in the fog, a truck ran through a stop sign and T-boned the Prince's car. Both Mr. and Mrs. Prince and Jason were killed and Janice was rushed to the hospital with serious but non-life threatening injuries. [*Real names not used*]

Marty continued with more details but it hit me at that moment that the leadership question had been answered. Marty, Mark and Ann were looking to us to lead the team through this crisis. The questions that haunted me at the start of the school year were going to be answered, and quickly. Other questions now needed to be asked. Would the team accept the four of us as leaders? What could I possibly offer to the team as a leader? I always saw myself as a worker bee, dealing with children, staff, and families, one on one, or in small groups. But these questions had to wait while the four of us sat down to begin planning. Now new questions began to arise. Questions not about myself, but about the team and how we needed to respond. It was as if the four of us looked at each other with a collective "we got this", and we began assuming the role of leaders.

The death of a family was a tragedy, but this family was very involved in their community, church, and school. Mr. Prince was a Boy Scout troop leader and an assistant baseball coach of Jason's team. Mrs. Prince was president of the middle school PTA, where Jason attended. Their house was where all the kids went to hang out. Jason was going to a baseball tournament as a member of a traveling team. Janice was just as involved in her high school, and Mrs. Prince made sure that she was available there as well. So many people and groups were impacted by the loss of this family. At that moment, I remember thinking how glad I was that Leslie, Donna and Claudette were with me to navigate this horrific task. All my training and reading about leaders, team leadership, and self-efficacy came to mind (Bandura, 1982; Crisis Prevention Institute, 2000; Everly, 2011; Katzenbach and Smith, 1993; Roth, 2015). I was moved in that first meeting of the Core Team how easily we fell into our respective leadership roles. The comfort with which we accepted the responsibility, not because we were individually great leaders, but collectively we had confidence in each other that the support and skill set would be there. We were ready to be flexible, and at the same time make decisions and take control. Through our ability to successfully communicate and collaborate with each other, we all moved forward, knowing and acknowledging that the task was going to be difficult.

Steps Forward

The four of us sat around the table and drew from our combined 60+ years' experience. We would need all of it. This was a multi-school and a community response. We spent most of Sunday, the next day, meeting with the primary, lead members of each school staff—principals of both the high school where Janice attended, and the middle school where Jason attended. We called in the school based crisis teams from both schools, and worked with them to prepare for their role in the coming days. Each member of the district response team was contacted. We decided that two main teams would be deployed, one to the high school and one to the middle school. This was a multi-issue response and the teams needed different preparation. The team at the middle school would be dealing largely with students grieving the loss of a classmate, while the team at the high school would be dealing with students coming to grips with the loss of life, but also their peer who lost her family and lived. Questions about what would happen to her, who she would live with, and her emotional status and needs would be the topic of group sessions. The age difference between the students was also considered. Both groups would view the situation from very different developmental perspectives.

I took the lead in the high school response and Leslie took the lead in the middle school. Monday came and we met with our respective building staffs in a 'stand up' meeting first thing in the morning. We planned to use our training in PREPaRE to make sure our communication was appropriate. PREPaRE workshop 1 indicates when communicating in a crisis, make sure to share facts of the situation in a way that:

- Conveys a sense of calm, confidence, and concern for wellbeing;
- minimizes heightened emotion;
- provides timely access to facts and minimizes rumors and gossip;
- respects the privacy and wishes of the immediate family;
- is attentive to possible issues of contagion;
- emphasizes safety, minimizes harm, and focuses on helping self or others.

The entire high school staff had been contacted by telephone over the weekend, so the news was not new, but no less difficult, especially as the reality hit when they all saw each other. In the meeting we restated the facts of the crisis and then told the staff how the day was going to proceed. The district level team members were introduced, and staff was told the process for having students come to us. The process and the "morning stand up" was no different than the 30+ meetings I had stood in prior to this one, except this time, I was the one standing up front. I was the one the administrator was looking to for guidance. The stress of Saturday, Sunday and now Monday morning was starting to build in my mind. I had a new appreciation for Jeff and what he had to manage. Secretly, I was looking around for him, because "care for the caregiver" was now much more than a phrase to me. The response and the need that day cannot be minimized. Nor can it be captured in these words on print. Such loss, so suddenly, leadership of the Boy Scouts, baseball teams, PTAs, community activities, all taken away by a foggy morning and a truck.

Our team handled the initial wave of students who came to the "Raider Room", the location where we did psychological triage (Brock et al., 2016). My team was responsive and immediately accepted my leadership. They moved with the precision that our team was known for, because of their preparation and ongoing training. Students moved steadily in and out of the Raider Room and small groups were worked with in break out rooms. I had just finished meeting with the administration at the high school to check on them, a role that I performed prior to the leadership position. In my experience, administrators are notorious for taking on too much to make sure their building gets what it needs, to the detriment of their own needs. They forget the simplest necessities, like using the bathroom or drinking water and eating. So I frequently checked on them starting the conversation with, "How's it going, have you been to the bathroom lately?" Often, the administrators almost need permission to take care of themselves. I would often follow that statement with permission, "You know I can sit in your office and monitor the phone while you go . . . and grab a bottle of water for yourself on the way back." While I was walking back to the triage room my cell phone rang. It was Leslie, asking me to come to the middle school because the distress there was overwhelming. At the time of the call there were 85 students in the library being used for triage, and just 12 or 13 responders. I told her I would be there and bring two members from the high school team since we were busy, but managing.

Care For The Caregiver—More Than A Statement

At this point in the response I was in *crisis mode*. My training and experience were all being utilized. Decisions were being made at a rapid pace, modifications to plans, moving of personnel, debriefing team members, working with administrators, and revising our response based on evolving needs and issues we had not anticipated. Issues such as parents coming to the school in large numbers, not to get their children, but because they needed to process the situation for themselves. The middle school seemed much needier than we anticipated. We underestimated the impact of the entire family on the school and the community. So, in *crisis mode*, I drove over to the middle school. Leslie met me in the front office. We met briefly with the principal in his office. Leslie and I debriefed each other on our responses. I asked how she was doing and how the team was handling it. She assured me that they were all hanging in, but the number and frequency of students flowing into the library was continuous. At this point, a team member came in announcing the two additional members had arrived from the high school and they needed to know how they should be utilized. It is interesting when you are in a leadership role how many requests and questions can come at you all at once. One minute you are debriefing with your colleague, the next you need to answer questions, or put out fires.

Leslie and I left the principal's office and headed down the administrative wing to the door that exited into the hallway, directly across from the library. I opened the door for Leslie and she went through, with me closely following. Exiting the library was a young seventh grade student with his head down. He glanced up and we made eye contact. He ran toward me throwing his arms around me and began sobbing. After a few seconds of holding him in my arms, he looked up at me and with tears streaming down his face said, "Dad, Jason was my science partner." Those words stopped me in my tracks. This crisis just took a serious *personal* turn for me. In all my preparation and work over the weekend and all day Monday, it never occurred to me that my youngest son, who attended the school, might know Jason. I found the answer in that tear-filled moment. My mind started going a hundred miles an hour. Where in my training did I learn how to handle a major crisis, as a leader of a crisis team, when my own son is intimately connected in a way that painfully moves him to such tears. Of course, I found no information in my memory of training for such a response. The truth of the matter is that crises, by definition, 1) are perceived as extremely negative (Carlson, 1997) and cause extreme physical and psychological pain (Brock et al., 2002) are uncontrollable, generating "feelings of helplessness, powerlessness, and entrapment" (APA, 2000); and 3) they often occur suddenly and without warning (APA, 2000).

And so I quickly found myself in a personal crisis within a larger school crisis. In an article written for *Principal Leadership* Journal, (DiRaddo & Brock, 2012), I outline a number of variables that put individuals at greater risk for traumatic reaction (Brock et al., 2016):

1. Physical proximity: the closer to the crisis event an individual is physically located, the greater the risk of psychological trauma.
2. Emotional proximity: individuals who have or had close relationships with the victim of a crisis have greater risk of trauma.
3. Internal personal vulnerability: having an avoidance coping style, poor self-regulation of emotion, or a history of trauma or mental illness increase the risk of trauma.
4. External personal vulnerability: having poor family resources, being socially isolated, or having a perceived lack of social support increase trauma risk.

5. Threat perceptions: adult reactions are an important influence on children's perception of risk and hence trauma risk.

Dealing with this very intense crisis situation with all of the facets of need, none of these risk factors were applicable to me. But now with the knowledge of my son's involvement, physical and emotional proximity as well as threat perception were real factors that I had to monitor and consider. I needed time to "reset", to "recoup", but that time, as is often the case during crises, was not afforded given the magnitude of the response. So the first thing I had to do was be a Dad. I took my son and we went back into the administrative wing where the principal allowed us to use his office. We were together only about ten minutes, but my son was able to regain composure—partially, I believe because I was able to maintain composure myself (threat perception). We talked about his relationship to Jason and what he remembered most. He laughed as he shared science lab stories with me. I found myself not even wanting to dip into my "therapist" role. Being Dad was enough . . . my son returned to class, and I assured him I would be in the building the rest of the day if he needed me. He did not return.

Leslie was the first through the door when my son returned to class. She wanted to know how I was and what she could do. We spent a little bit of time debriefing the new situation, my crisis within the greater crisis. More than one tear fell from both of us during our debrief. She was concerned that I might not be able to continue at the middle school and suggested I return to the high school. I declined. There was something about seeing my son's reaction and his rebound that gave me additional strength. I am not suggesting that he was not on my mind, or that he was "fine", but I felt empowered to move forward and help as many of the students and families as I could. Little did I know that meant three days of responding, including night meetings with parents. The rest of Monday was spent seeing children, checking on the administrators, teachers, and staff. We broke for lunch with the whole team and "closed the library" for 45 minutes—a necessary step so we could debrief and allow the team some time to talk and care for their own needs. I have always found this step difficult, but necessary. It is hard to not allow kids in need access to your help, but without a break and a reset, the effectiveness of the responders is compromised.

RECOVERY

Parent Meeting and Caregiver Training

Monday and Tuesday afforded the children an opportunity to come and talk as they needed. On Wednesday, we began asking students to get passes and we began considering referrals as needed. We also determined that a parent meeting in the evening was necessary due to continued parental inquiries.

Parents needed to know what they could do. They also needed to gain an understanding of what they might see from their son or daughter in the next few weeks. They needed to hear that much of these children's reactions were typical, and they needed to give them reassurance of their personal safety and time to recover. So the Core Team spent a good bit of Wednesday preparing for the meeting. We also were able to line up outside resources and agencies willing to provide information during the meeting about services they provided. As the Core Team entered the meeting, the middle school library was packed with concerned parents. Information was made readily available for them. Part of the information we shared was that their children were reacting in many ways, but that there was no one "right way" to respond. We attempted to normalize their children's reaction to this horrific situation, but suggested they look for excessiveness in any one area: sleeping, eating, isolation, etc., as "red flags" that might indicate a need for more help.

We talked about media and technology. Yes, this is the first time I have mentioned technology, but of course it played a huge role in this crisis. Cell phones and social media were ablaze from Saturday morning on. Talking to my son in middle school and my daughter in the high school, they shared the circulating rumors and comments. This is an area that always needs attention and monitoring when responding to this kind of far reaching crisis. During the parent meeting I again reflected, as each member of the Core Team presented their respective pieces of information, that joint leadership with competent co-leaders was truly a blessing. We were each able to support one another, as well as jointly field questions that gave clearer, more pointed answers than if one of us was leading alone. For example, I recall that Leslie did an amazing job fielding and handling technology/social media questions.

LESSONS LEARNED AND REFLECTIONS

So we "handled it", but at what cost? The four of us sat together on Friday, and looked at each other. The cost of leadership for those four days was evident in our tired eyes and drooping posture. We began to debrief again and examine the entire week. What would our next steps be? We knew it was not over. We knew there was more we would have to deal with, so what should we be prepared for? We began planning again. How to handle the day of the funeral was the first topic. As time went on we met about support for the surviving daughter, for Jason's baseball team, for the Boy Scout troop, the first year anniversary, and next year the celebration of Janice's graduation from high school. You see, when you have a crisis like this one, the first lesson learned is that it is never 'done'. The impact is real and lasting. I found myself dealing with my own issues regarding the response and my son for several weeks after the response was over. I had to make a conscious effort to follow my own words given to my team as we wrapped up:

- Get enough sleep and rest.
- Make sure to eat and exercise.
- Follow your religious practices and be around your support group.
- Talk about your reactions, but don't dwell on them.
- Get help if you need it.
- Be honest in monitoring yourself and get relief if you need it (Brock et. al., 2016).

These were all things I suggested to my team, in a meeting we held two weeks after our response was completed. But could I take my own advice? The answer was *yes*, but only with the help of my support system—the Core Team. The bond that was forged in that first leadership event continues to this day. Other crisis situations have arisen each year and we have met those challenges as a Core Team as well, but we always go back to our first response together and reflect on the lessons that we learned during those long days and evenings. We helped each other process this first response without fear, admitting that we needed to process how we were dealing with it. This processing with each other, brings me back to Jeff Roth's chapter on "Leadership and Teamwork" (2015). We met the three areas that he spells out as important qualities of leadership:

1. Comfort with taking and sharing crisis leadership;
2. supporting the mission before, during, and after the crisis, and;
3. supporting resilience among affected persons and team members.

What Was the Take Away?

In conclusion, the lessons I learned in my training, whether PREPaRE or more powerfully "on the job", are all lessons that should be kept in mind when response and leadership are needed:

1. Be flexible—don't be afraid to change the plan.
2. Listen—not only to your peers, but to that voice inside that tells you how you should respond.
3. Listen—to the people in crisis. They will let you know in their words and actions what steps you need to take to be helpful.
4. 'Stick to the script'—sounds contrary to be flexible, but stick to the script means, make sure that you state the facts, don't "ad-lib" when talking to groups of people. There is no harm in saying, "I don't know that answer right now, but I will find out and let you know."
5. Do not be afraid to accept help. When you spend your time as a caregiver, eventually you have nothing left to give. So take the time to replenish and re-coup your physical and mental energy unashamedly, knowing it is necessary for being able to help others.

6.   Don't assume you are aware of the culture of the building you are working in, just because the group you are working with is in your school district. Each grade level, each school, has it's own set of norms, standards, cultural and developmental considerations, and beliefs. For example, in the same school district, I have worked in one high school where the students would not come together as one big group but rather as smaller isolated groups, and in another high school where they wanted to bond together as a big group. Responding to each school takes a different course of action.

7.   Look out for the other caregivers. Make sure to check on those who are doing the hard work. Intervene if you have to, to give them a break.

8.   FINALLY, look out for yourself so you can continue to care for others.

REFERENCES

American Psychiatric Association. (2000). *Diagnostic and statistical manual of mental disorders* (DSM-IV-TR) (4th ed.). Washington, DC: APA.

Bandura, A. (1982). Self-efficacy mechanism in human agency. *American Psychologist,* 37, 122–147.

Brock, S.E. (2011). *PREPaRE Workshop #2: Crisis intervention and recovery: The roles of school-based mental health professionals.* (2nd ed.). Bethesda, MD: National Association of School Psychologists.

Brock, S.E., Lazarus, P.J., & Jimerson S. R. (eds) (2002). *Best Practices in School Crisis Prevention and Intervention.* Bethesda, MD: National Association of School Psychologists.

Brock, S.E., Nickerson, A.B., Louvar Reeves, M.A., Conolly, C.A., Jimerson, S.R., Persce, R.C., & Lazzaro, B.R. (2016). *School crisis prevention and intervention: The PREPaRE model.* (2nd ed.). Bethesda, MD: National Association of School Psychologists.

Carlson, E.B. (1997). *Trauma assessments: A clinician's guide.* New York: Guilford Press.

Covey, S.R., Covey, S., Summers, M., & Hatch, D.K. (2014). *The leader in me: How schools around the world are inspiring greatness, one child at a time.* (2nd ed.). New York: Simon & Schuster.

Creamer, T.L., & Liddle, B.J. (2005). Secondary traumatic stress among mental health workers responding to the September 11 attacks. *Journal of Traumatic Stress,* 18, 89–96.

Crisis Prevention Institute. (2000). *Nonviolent Physical Crisis Intervention Training Manual.* Milwaukee, WI: Crisis Prevention Institute.

DiRaddo, J.D., & Brock, S.E. (2012, May). Is it a crisis? *Principal Leadership, 12(9),* 12–16.

Everly, Jr., G.S. (2011). *Fostering human resilience in crisis: A primer on psychological body armor and psychological first aid and resilient leadership.* Ellicott City, MD: Chevron.

Feinberg, T., Pfohl, W., & Cowan, K. (2004). Crisis: Tips for Caregivers. In A. Canter, L. Paige, M. Roth, I. Romaro, & S. Carroll (Eds.), *Helping children at home and school II: Handouts for families and educators.* Bethesda, MD: National Association of School Psychologists.

Katzenbach, J.R., & Smith, D.K. (1993). *The wisdom of teams: Creating the high-performance organization*. New York: Harper Collins.

Reeves, M. (2006). National Association of School Psychologist Convention, PREPaRE Pilot project training, Anaheim California.

Reeves, M.A., Nickerson, A.B., Conolly-Wilson, C., Lazzaro, B., Susan, M.K., Pesce, R.C., & Jimerson, S.R. (2011). PREPaRE Workshop # 1: *Crisis prevention and preparedness: Comprehensive school safety planning*. (2nd ed.). Bethesda, MD: National Association of School Psychologists.

Roth, J.C. (2015). *School crisis response: Reflections of a team leader*. Wilmington, DE: Hickory Run Press.

US Department of Homeland Security (US DHS). (2008). *National incident management system*. Washington, D.C.: Author. Retrieved from www.fema.gov/pdf/nims/nims_doc_full.pdf

2 Remembering Carlos
Professional and Personal Reflections

Andrew Vengrove

INTRODUCTION

Having completed over 30 years as a school psychologist, I have often reflected on the following incident as it deeply impacted me, both professionally and personally. As background, in 1989 a colleague and I developed a system-wide intervention plan for suicide, sudden death, and catastrophic school events in response to the lack of structure when counselors and psychologists were asked to assist without a plan. Over the years, we have provided interventions for deaths of students, faculty, and community members as well as incidents related to weather and national disasters like the World Trade Center. While these incidents were notable, the one described below stands out, as reminders of "Carlos" and the lessons learned often re-emerge.

Personal Impact

There are several reasons for the impact of this event on me. First, I knew Carlos and a member of his family when he was in middle school. Having personal connections to the deceased can be tricky. I had to be aware of my emotions during the whole process. Yet, the personal connection was also helpful in bonding with others over the tragic loss experienced by students, staff, and the community. Second, although the crisis team was unprepared for this unexpected event, I came away with appreciation for how well we worked together and trusted one another. We were flexible in our thoughts and actions, and avoided becoming emotionally immobilized by 'catching the crisis.' Third, I was impacted by the number of students and families affected by Carlos' death and the pain emanating from the initial gathering of those who appeared in shock, grieving the immediate loss of their friend and community member. Fourth, I teach a course at a local university entitled, "Crisis Intervention in the Schools and the Community," where I have had a number of counseling students who attended the high school at the time of Carlos' death. I have come to realize that Carlos is very much alive in the hearts and minds of these young graduate students, seeing and feeling their emotions as we discuss lessons learned from system-wide school interventions. I admit that it felt good to hear that the efforts of our crisis team were remembered and appreciated. Last, I gained much respect for the large numbers of students involved during the initial interventions, as we learned to work and grieve together while being respectful of all involved.

DESCRIPTION OF RESPONSE

Community Impact

On a Sunday in November at 12:13 a.m., a tragic car accident in a Massachusetts town killed a popular high school student. On that evening, the local news reported that five high school students were travelling in a car at a high speed when the driver lost control and crashed into a tree. One student was killed, one seriously injured, and the three others, including the driver, were treated and released that evening.

In order to understand the overwhelming response by the community, it helps to be introduced to the young student who lost his life. Carlos was a unique person who had an impact on many people. He was 16 years old when he died. He was unique because he was not part of any one group or "clique" often found in high schools. He seemed a part of all the students and had friendships that cut across all boundaries. Carlos was a member of the high school's gymnastics team that won the Western Massachusetts Championship ten days before the crash. He was a member of the wrestling team and played percussion in the school band. Carlos was also a drummer in a Christian rock band that was preparing for a tour in the coming months. Many people knew Carlos and were affected by the death of this young person who was so full of energy and life. Often when tragedies occur, local public schools are a place where high school students and their families congregate for a sense of community and support.

Initial Crisis Team Briefing

As stated in the district's system-wide crisis intervention plan, the principal received the initial phone call from the local Police Chief informing him of the accident. Knowing the complexity of the situation, the high school principal called each member of the crisis team at home and asked that they assemble at the school on Sunday afternoon to plan for support in the following days. The school crisis team included the principal, vice principals, school nurse, counselors, school social workers, school psychologists, special education supervisor, head custodian, and the school safety officer. The initial purpose of the meeting was to do the following:

1. Sort out the facts of the car accident (i.e. fatalities, injuries, driver, etc.).
2. Identify the various school groups associated with Carlos (i.e. band, athletics, etc.) and with the other students involved in the crash.
3. Identify students closest to Carlos or who may be at risk after hearing the news.
4. Identify a crisis team member to contact Carlos' family.
5. Identify faculty members who might be most affected by his death.
6. Identify possible support within the community.
7. Discuss concerns about the 17-year old driver, who was also a student at the school.
8. Locate and prepare counseling areas for students on the following Monday.

Coping with Unexpected Demand

Part of crisis work is to be prepared for the unexpected—the "unexpected" occurred while the crisis team members were preparing for Monday's school-wide interventions. Given the available technology—cell phones, text messaging, email, word travels quickly to many people. School counselors had some communication with at-risk students who were aware the crisis team was meeting on Sunday afternoon at the high school. The meeting "switched gears" when to the team's surprise, they realized that hundreds of students, families, and community members were arriving at the high school seeking a sense of community and help for those in need. While the high school crisis team had dealt with significant past crises, the number of people who kept arriving in a state of shock and grieving was staggering and unexpected. From a clinical perspective, it was important to monitor our own emotional state in order to keep focused and insure that, as members of the crisis team, we had the proper mindset to work effectively with the immediate situation. For example, Vengrove and Rice (2009) note:

> From our experience working with people and with larger systems in crisis, creating a structure around the perceived or real internal and/or external "chaos" is one of the keys to success in crisis work. In order to be an effective crisis worker, we contend that crisis workers must maintain their own equilibrium in that they must enter the crisis situation with the proper mindset. We must be able to maintain our professional stance while also being able to enter into chaotic, emotionally charged, and potentially dangerous situations . . . Simply stated, the first "rule" of crisis intervention is to not "catch the crisis".
>
> (pp. 275–276)

As approximately 400–500 students, families, and community members assembled in the cafeteria, the focus of the crisis team changed from planning for Monday to planning for the immediate. This meant securing a large area and having water, tissues and staff available and prepared to administer psychological first aid. Psychological first aid is usually, "short-term and used to help clients reestablish immediate coping" (Myer, 2001, p. 26). Maintaining our equilibrium as responders was crucial, yet challenging as the level of pain and emotions emitted by students and their families was overwhelming. While members of the crisis team made their way through the crowd to check in with various groups that formed, parents often requested assistance for their child, expressed concern for another student, or asked about plans for the coming week. The principal addressed the crowd, stating the facts and asking for a moment of silence. The principal gave a brief overview of the services that would be available during the next school day. The counseling coordinator next addressed the crowd with self-care tips and places to get immediate help from crisis team members at the meeting or in the community at large. Last, I addressed the crowd as a school psychologist, system-wide coordinator of the crisis program, and one who knew Carlos, sharing some personal comments about him, and emphasizing adaptive rather than maladaptive ways to deal with grief. Many students seemed to appreciate the personal comments about Carlos. It seemed important to talk directly about him and the person he was rather than focus only on the palpable grief within the cafeteria. In the midst of their grief, many students thanked me for my comments. Several organized "break-out" groups formed in the cafeteria and other parts of the school building—sports and music-related groups to which Carlos belonged. These groups were facilitated by their teacher sponsors or school counselors.

Crisis Team Debriefing and Continued Preparation

At the conclusion of the meeting, the crisis team met again to debrief and continue planning for the next day. In addition to the tasks noted earlier, team members:

1. Generated class rosters to identify all of the classes that Carlos attended.
2. Discussed the impact on other district schools, especially family and friends attending those schools.
3. Notified all high school teachers of the death and instructed them to attend an emergency faculty meeting early Monday morning.
4. Prepared sign-up sheets for the school's counseling center on Monday. As learned from previous crises, all students receiving assistance were asked to "sign-in" with their name and home phone number. Every student's parent or guardian was called by a team member and informed that their child sought services. This created a beneficial, positive link with families. It allowed parents to ask questions about what to expect, and team members could convey concerns and provide information about referral, if needed.

5.   Last, the importance of de-briefing with fellow colleagues was emphasized to help responders manage their emotions and not "catch the crisis."

Faculty Meeting and Support

Before school on Monday morning, the entire faculty and all members of the system-wide crisis team met for an emergency meeting. The purpose was to:

1.   Update faculty on Sunday's events and facts that was learned over-night, including the status of the students who were injured and still hospitalized.
2.   Discuss procedures for the day including:

   • identifying crisis team members and roles throughout the day;
   • providing information to teachers on "what to expect" and suggestions on how to conduct their classes and handle students in need;
   • identifying the "Counseling Center" set up in the school auditorium to accommodate the anticipated large number of students in need;
   • identifying ways that teachers could get support if they were overwhelmed by the tragedy and the day's events;
   • highlighting that flexibility within the structure of the day would be essential as the reactions of grieving students can vary. For example, it has been suggested that while large numbers of students will be directly affected by the loss, others may not be close to the deceased, yet seek help for their own issues with loss and mortality (Swihart, Silliman, & McNeil, 1992). Many students may be in a state of shock, experiencing a sense of disequilibrium and not able to perform to the best of their ability. Given the definition of disequilibrium, "Lack or destruction of emotional stability, balance, or poise . . ." (James & Gilliland, 2013, p.50), it is important to acknowledge that the equilibrium of the building and community could be affected for the present and for days and weeks ahead.

Identifying, Intervening with, and Empowering Students Needing Support

Upon arrival to school, students assembled in their homerooms where the principal addressed the faculty and student body by acknowledging the loss, having a moment of silence, and providing information for those students in need of help. At the conclusion of homeroom, a large number of students filed into the auditorium, each signing in as they entered. Members of the school crisis team, including additional counselors and psychologists from across the district were disbursed throughout the auditorium attending to large and small groups, and individual students in need. Activities included:

1.   Informal counseling with groups and individuals in the auditorium and in smaller areas within the building (i.e. administrative, health and counselor offices).
2.   Attending especially to vulnerable students at higher risk—those in need of individual attention from counselors including contact with parents or guardians.
3.   Distributing large pads of paper and markers to allow students to draw or express their feelings, messages, and/or memories. The principal allowed these to be displayed in the auditorium throughout the morning. They would later be presented to Carlos' family.
4.   As a school psychologist who knew Carlos, I addressed the students acknowledging Carlos, educating about the grieving process including adaptive and maladaptive ways of handling grief, and identifying ways to seek help. I praised the manner in which such a large group of students were able to assemble in such a respectful manner.
5.   Most impressive and memorable was that, within 12 hours, Carlos' closest friends put together an emotional multimedia slide show that included pictures of Carlos accompanied by music. The students brought the slide show to school and negotiated with the crisis team for permission to play it periodically throughout the morning. It seemed to help the close friends to collectively focus their grief. The slide show was quite moving and helped those who chose to view it experience both laughter and sadness as the photos (and memories) highlighted the many sides of Carlos. All pictures and music were respectful and appropriate for the school setting. Viewing the show was voluntary, and counselors were available to help students process reactions afterward.

MOVEMENT TOWARD RECOVERY

Normal Routine with Ongoing Support

These activities took place for most of the morning and early afternoon. Constant communication and assessment by members of the crisis team was evident as the focus often shifted from individuals or groups of students, to teachers and the larger school environment, and to interactions with parents and other community members. At one point there was a sense that students needed permission to move on with their day—gain some sense of normalcy. This proved to be correct as there was no resistance when students were eventually asked to try continuing their normal routine, with counselors available if needed.

A debriefing session with crisis team members occurred at the conclusion of the day, including plans for the next day. There was only one day of school left before the Thanksgiving vacation. Counselors continued to be available for students in need during the following day and weeks afterward. School personnel stayed in contact with the grieving family. Some members of the crisis team, including the principal, attended Carlos' funeral. With the help of students and faculty, the high school established a yearly music scholarship award in Carlos' name.

LESSONS LEARNED AND REFLECTIONS

In retrospect, there were some key lessons learned from this experience:

1. When severe crises occur, schools may become a gathering place for the community. While crises are difficult situations for those affected, they can also be an opportunity to grow as individuals and as a community (James & Gilliland, 2013; Kanel, 2007).
2. Never underestimate the power of technology as word travels quickly and this can affect planning and the need for immediate interventions.
3. Flexibility within the school-wide system shows respect for individual differences in reactions to grief. For example, the large body of affected students felt they had a voice in the process, reducing negative reactions expressed in the context of the interventions.
4. During a crisis situation, "assessment" must be constant (James & Gilliland, 2013), to help determine individual and system-wide support needs. Within the assessment process, communication among team members is a key to establish the proper mindset for crisis work. Communication and coordination helps each member work within their pre-designated role to help others and create an organized structure from a potentially disorganized and chaotic situation.
5. Having a personal connection with the deceased while providing psychological first aid requires full awareness of the range of emotions experienced. Monitoring and regulating feelings helps to not "catch the crisis" and to be an effective team member. Still, one can model feelings and use the personal connection to bond with the community and speak sensitively about the essence of the deceased.

CONCLUSION AND PERSONAL REFLECTIONS

When working in schools, it is inevitable that experiences involving death, pain, and grief will occur and we will be called upon to assist. Effective crisis management is critically important for a school system to support student needs when there is a death. Crisis management needs are often immediate, and involve providing effective interventions to potentially large numbers of people within a short amount of time. The effects of these events can be long-lasting, and impact not only friends, family, and school community, but also members of crisis response teams. While these events have been exceedingly difficult, I am aware of also being deeply affected in a positive way. I have reflected upon "lessons learned" and applied these lessons to improve our approach to crisis work, and to teach crisis intervention skills to graduate students. Most important, effective crisis response and interventions support the constructive expression of grief, remembrance, and meaning for students like Carlos, who remains in our hearts and minds.

REFERENCES

James, R.K. & Gilliland, B.E. (2013). *Crisis intervention strategies* (7th ed.). Belmont, CA: Brooks/Cole.

Kanel, K. (2007). *A guide to crisis intervention* (3rd ed.). Belmont, CA: Brooks/Cole.

Myer, R.A. (2001). *Assessment for crisis intervention: A triage assessment model.* Belmont, CA: Brooks/Cole.

Swihart J., Silliman, B., & McNeil, J. (1992). Death of a student: Implications for secondary school counselors. *The School Counselor, 40,* 55–58.

Vengrove, A. & Rice, S. (2009). Developing a professional mindset: Individual and organizational crisis intervention strategies. In Young, N. & Michael, C. (eds), *Counseling with confidence: From pre-service to professional practice.* Amherst, MA: Synthesis Center, Inc.

Section 2

School Crisis Response to Death of a Student

INTRODUCTION

Children and adolescents are expected to live long lives, so the death of a peer is especially sad and shocking. Identifying with a deceased classmate adds a layer of fear and vulnerability to grieving.

A school's response to the death of a student influences the process of recovery. It is imperative to support healthy grieving and coping, while minimizing emotional, behavioral, and academic problems. Effective response to tragedy can enhance trust among students and staff, promote resilience, and strengthen the school community.

Prevalence

Mortality data for 5–14 year olds indicated 12.7 deaths per 100,000 population, and for 15–19 year olds, 45.5 deaths per 100,00 population. Leading causes of death for 5–14 year olds were accidents, cancer, and suicide, while leading causes of death for 15–19 year olds were accidents, suicide, and homicide (National Center for Health Statistics, 2016). Clearly, most school-age deaths are sudden, rather than anticipated. Sudden student deaths have been estimated to occur more than once every week in the United States (National Center for Education Statistics, 2009).

Context of the Death

The circumstances of a student death affect the impact on the school community. While the death of any classmate is tragic, when the deceased is well known and admired, the impact can be far-reaching. When the death is sudden, shock and disbelief often leaves no time for survivors to anticipate or prepare for intense emotions that follow. When the death occurs on the school campus, emotional trauma, grief, and stress reactions can be intensified, especially for witnesses.

Grief

A supportive environment, stable adult figures, and early intervention can be helpful for grieving youth (Worden, 2009). Bereavement is a normal reaction and the expression of grief includes thoughts, emotions, physical and behavioral reactions. Mourning a death reflects cultural traditions and comforting rituals. The most common pattern of bereavement is *resilience*—a period of grief followed by adjustment to the reality of loss, and moving forward to meet the demands of life.

Sudden, unexpected, untimely deaths are more likely to create problems such as *complicated grief*. While sadness and other typical reactions are considered normal and adaptive, complicated grief can include more intense reactions, stronger emotions, intrusive disturbing memories, and difficulty with daily functioning. Early intervention is critical for students experiencing complicated grief.

Since bereaved students may experience various reactions during developmental transitions, grief can be re-experienced in different ways over time. School staff who are sensitive to students' reactions and long term needs after the death of a classmate, can make certain that appropriate ongoing support is available (Jimerson et al., 2012a).

PREVENTION AND PREPAREDNESS

Promoting Children's Resilience

Promoting children's resilience begins before a death occurs, starting with a supportive home and school climate that is responsive to needs and feelings. If a death is anticipated, children can be prepared for what to expect, for the funeral, and for typical emotions. While the norm is resilience and recovery, evaluation of risk for emotional trauma can help identify those needing special attention. When adults support children and adolescents coping with a death, they can model, practice, and teach the principles of resilience. They can teach young people about the grieving process and normalize fearful reactions. Poland and colleagues (2014) summarize strategies for promoting resilience.

Preparedness

Proactive planning can minimize emotional trauma experienced after the death of a student. Preventing a sudden, unexpected death may not be possible, but preparing for such an event is possible by designing comprehensive, flexible response plans that address a wide range of potential, unexpected crises (Jimerson et al., 2012b).

Suggestions to plan for preparedness:

- Establish coordinated school, district, and community crisis response teams.
- Prepare response plans that are both specific and flexible.
- Train teachers, administrators, and support staff about grief reactions, crisis interventions, and developmental and cultural competence.
- Provide procedures and strategies for school staff to facilitate student coping.
- Develop tracking systems to document student-related contacts during crises.
- Coordinate services with community organizations and resources.
- Provide parent workshops and consultation publicized at PTA meetings, in parent newsletters, and on "back to school night" flyers.
- Prepare follow-up plans after the initial crisis response.
- Consider that sudden deaths may require *immediate* response for those affected, and address this consideration when developing response plans.

Developmental, Cultural, and Special Needs Competency

The effectiveness of interventions is influenced by student development, cultural diversity, and special needs. While most children understand aspects of the death concept by age 10, developmental factors influence their reactions and needs. For example, elementary students (age 6–12) are more likely to fear that a supernatural death may stalk them and loved ones, perceive death as punishment for bad behavior, or engage in guilt-inducing *magical thinking* that an unrelated thought or action caused the death. Middle and high school students (age 12–18) generally realize that death is permanent, but often philosophize about death, fantasize about their demise, and feel invincible, which leads to reckless behaviors. Because self-esteem through friendships becomes increasingly salient toward adolescence, it is vital to give special attention to friends and peers of a deceased student (Jimerson et al., 2012a).

Schools reflect our multicultural society and require a comprehensive response plan sensitive to the diverse customs of their ethnic and racial groups, and connected with community resources. Various cultures have different attitudes, norms, and ways of coping with death. Culturally mediated reactions can vary from dramatic and animated to stoic, so that understanding these differences is especially relevant when evaluating emotional trauma (Ortiz & Voutsinas, 2012).

Students with physical, emotional, cognitive, or behavioral disabilities may have more difficulty processing and coping with adversity. Students who have special needs may require more intensive support and careful monitoring after the sudden death of a classmate (Jimerson et al., 2012a). When working with students having special needs, it is important to respect each student's dignity and recognize their strengths and abilities.

Training and Consultation

PREPaRE model

The National Association of School Psychologists (NASP) provides a comprehensive school crisis response model, PREPaRE, which is easily adapted to a variety of traumatic events, including student death (Brock et al., 2016). It is highly recommended that school crisis response teams be trained in PREPaRE.

Trained educators as protective factors

Teachers in daily contact with their students can be a protective factor, identifying and responding to students struggling with grief (Candelaria, 2013; Rossen & Cowan, 2013; Schonfeld & Quackenbush, 2010). Workshops and training modules with consultative follow-up can provide teachers, administrators, support staff, and other caregivers with the knowledge, skills, and confidence to support grieving students after a tragic event. Suggested training modules include "recognizing typical and complicated student grief reactions (Heath & Cole, 2012), and topics developed by Poland and colleagues (2014), including "strategies to help bereaved students cope and recover", "classroom activities to support resilience of grieving students", and "strategies for parents to support grieving children and adolescents". Modules can also be available online.

Suggestions for teachers and staff to support bereaved students:

- Educate teachers and families about supporting normal, healthy grief, and about maladaptive, complicated grief reactions that signal need for treatment referral.
- Tell affected classes about their classmate's death without disturbing details, and inform about available counseling resources, if needed.
- A script for informing students should be available to teachers as well as in-class support from a crisis responder if requested.
- Let young students know that death is not contagious—it cannot be "caught".
- Support and escort students seeking crisis counseling resources.
- Acknowledge the intense emotions evoked by a classmate's death—calmly model expression of your feelings, but never pressure students to talk.
- Listen nonjudgmentally, acknowledging feelings in a calm manner—be patient with repeated questions, and ready to repeatedly reassure safety.
- Answer questions with brief, honest responses that dispel rumors—focus on student reactions and not circumstances of the death.

- Help normalize a range of *typical* feelings and educate about grief, but connect students exhibiting maladaptive reactions to crisis counseling.
- Encourage adaptive coping strategies, especially interaction with social and family support systems (club, team, faith).
- Teach coping strategies such as talking with a trusted adult, exercise, mindfulness, deep breathing, and progressive relaxation to manage stress and intense feelings.
- Be sensitive to cultural differences in the way students and families express grief.
- Students with disabilities may need more support to understand and cope.
- Provide opportunities to express feelings—use teachable moments to address misdirected anger and promote understanding that not everyone feels the same.
- Resume normal school routine as soon as possible, but maintain support for those grieving and adjust academic expectations if needed.
- Those severely affected should be referred for individual attention, or to small groups with others who are similarly affected.
- When an entire classroom is similarly affected, consider a structured approach like a *classroom meeting, psychoeducational group,* or *classroom-based crisis intervention* led by trained co-facilitators (Brock et al., 2016).
- Be prepared for students who had conflict with the deceased to discuss unpleasant emotions, including anger or magical thinking and guilt.
- Assess the meaning of the death for each student—especially those who experienced a recent death or have a history of emotional problems.
- Expect regressive behaviors from early elementary students, including fear of being alone, fear of death, and somatic complaints.
- Avoid trying to quickly soothe feelings—instead, reflect feelings and empathize.
- Bibliotherapy—the use of selected books and stories can help the bereaved understand, discuss, and cope with a death.
- Educate about grief, encouraging students to be supportive of one another—high school students may show empathy through physical contact such as hugging.
- Facilitate adolescent peer group support, allowing students to share memories of the deceased, find meaning, and engage in life-affirming activities.
- Provide opportunities for constructive activities and self-expression such as writing, drawing, music, or creating a memory book for the grieving family.
- Avoid imposing your own religious or personal beliefs on others.
- Do not immediately remove the deceased student's desk or chair—perhaps wait until after the funeral.
- Continue connecting with students affected by the death to offer support, but without imposing when support is not needed—those expressing

intense emotions, withdrawal, or extreme behavior should be referred for support.
- Monitor progress toward recovery, being vigilant for symptoms of distress weeks or months after the death.
- Recognize that anniversaries, holidays, and birthdays may be difficult, and students may need extra support during these times.

Sources: Adapted from Fernandez et al., 2015;
Jimerson et al., 2012b; Poland et al., 2014

Social Media as a Challenge and a Support

While ideally there will be time for the response team to plan and meet with faculty before proceeding with interventions, communication through cell phones, texting, and social media may necessitate immediate response for students and the school community. This requires advance preparation for the impact of technology.

A group of students and school staff can work together to monitor social networking sites and social media. The school website and other media sites can be used as part of the response by providing *safe messaging* and information about school and community resources (Erbacher et al., 2015).

RESPONSE

Verify the Facts

- Principal should immediately verify the report of death with authorities and/or the family to be certain it is correct.
- Once confirmed, principal should inform the district superintendent and teachers, if possible, personally informing teachers who worked with the deceased student.
- Notify other schools which have students affected by the death.
- Brief office administrative staff and provide script for response to inquiries.

Mobilize the Crisis Response Team—Determine Level of Response

- See book **Introduction**, *Essential Elements* . . . Mobilization: Initial phase . . .
- Provide calm, coordinated response appropriate to the needs of the incident.
- Determining resources needed after the death of a student, consider 1) well known and admired deceased student generates more emotional trauma, 2) sudden, unexpected death generates more emotional trauma,

3) death on school campus with witnesses is especially traumatic, 4) previous traumatic events or deaths complicate coping with present death, and 5) purposeful act by a known perpetrator intensifies emotional trauma.

Initial Briefing and Crisis Team Planning

- See book "Introduction", *Essential Elements* . . . Crisis Team Planning . . .

Notify School Staff and Meet to Plan Student and Staff Support

- See book "Introduction", *Essential Elements* . . . School Staff Meeting.
- Stabilize and support teachers and staff, providing helpful information and counseling resources as needed.
- Provide classroom script, coordinating and carefully informing students of the death and escorting to counseling resources if needed.
- Provide in-class mental health support if requested by teacher.
- Consider in-class mental health support that follows the schedule of the deceased student.
- Establish location (often the library) to conduct triage and interventions, and *safe rooms* for students or staff to access support.

Notify Parents and Community about the Death and Resources for Support

- Contact grieving family, express sympathy, and determine wishes regarding public or private funeral.
- Provide letter to families with helpful information and counseling resources.
- Provide information about how to support students and access counseling resources through the school website and other media.
- Consider extending school hours, including weekend if needed, to provide counseling and partner with community agencies to provide services.

Plan to Manage the Media

- See book "Introduction", Media Plans and Cooperation.

Conduct Triage to Identify and Support Those Most At Risk

- See book "Introduction", *Essential Elements* . . . primary, secondary and tertiary triage, and variables for evaluating risk.
- Be especially supportive of those in close *emotional proximity* (family, friends, classmates, teachers, teammates) to the deceased student, *physical*.

proximity (witnesses to the death), those having *personal vulnerability* (history of trauma, loss, special needs, mental illness, lack of support systems), or showing *warning signs* (maladaptive coping, suicidal thoughts).

- Document all student and parent contacts, informing parents/guardians of concerns, ways to be supportive, and resources for referral if needed.
- Notify other schools having students who may be affected.

Source: Adapted from Brock et al., 2016

Conduct Interventions that Match Degree of Need

- See book **Introduction**, *Essential Elements* . . . Crisis Interventions.
- Provide honest reassurance of safety and security.
- Reestablish social support systems—family, friends, faith, and school.
- Provide range of tiered interventions appropriate to severity of needs—individual, small homogeneous group, or classroom-based.
- Interventions can be developed at the *universal* level (Tier 1), addressing all students school-wide, at the *selected* level (Tier 2), targeting students identified as struggling and needing more attention, and at the *indicated* level (Tier 3), for students having severe reactions requiring more intensive treatment (Brock et al., 2016; Gutkin, 2012; Meyers et al., 2012).
- Do not impose interventions that are not needed or do not match a student's degree of need—facilitate access to naturally occurring support systems.
- Consider developmental, cultural, and special needs to understand reactions and provide appropriate interventions.
- While intervening, conduct *secondary triage* to further identify those at risk, and ask students about peers who may need more attention.
- Peer group is a powerful resource for teenagers, but adult supervision is important to guide toward constructive, life-affirming activities.

Conduct Crisis Team and Staff Debriefings Throughout the Response

- See book "Introduction", *Essential Elements* . . . Crisis Team Debriefings . . .
- School staff appreciates being kept informed through meetings, memos, emails, and supportive personal contacts that enable care for students.

RECOVERY

See book "Introduction", *Essential Elements* . . . Disengagement: Final Phase of Response, which details suggestions for planning school memorials, the need for *tertiary triage* and follow-up, and response team debriefing.

- Funeral and memorial services often mark a return to normal routine with support as needed.
- Crisis response team disengagement must leave students in the hands of recovering school personnel who are capable of caring for them.
- *Tertiary triage* during crisis team disengagement should involve careful consideration of the need for continued monitoring, support, and/or treatment for those most severely affected.
- School staff should be aware of possible delayed stress reactions and depression, in order to more effectively monitor students and refer for risk assessment if needed—after severe emotional trauma, the need to monitor reactions may extend for months and years.
- Crisis team debriefing is critical for responder peer support, to process response effectiveness, and facilitate stress management according to individual need.

REFERENCES

Brock, S.E., Nickerson, A.B., Louvar Reeves, M.A., Conolly, C.A., Jimerson, S.R., Persce, R.C., & Lazzaro, B.R. (2016). *School crisis prevention and intervention: The PREPaRE model*. (2nd ed.). Bethesda, MD: National Association of School Psychologists.

Candelaria, A.M. (2013). Examining Kentucky teachers' encounters with grieving students: A mixed methods study. *Theses and Dissertations—Educational, School, and Counseling Psychology*. Paper 14.

Erbacher, T.A., Singer, J.B., & Poland, S. (2015). *Suicide in schools: A practitioner's guide to multi-level prevention, assessment, intervention and postvention*. New York: Routledge.

Fernandez, B., Comerchero, V.A., Brown, J.A., & Woahn, C. (2015). *Addressing grief: Tips for teachers and administrators*. Bethesda, MD: National Association of School Psychologists.

Gutkin, T.B. (2012). Ecological psychology: Replacing the medical model paradigm for school-based psychological and psychoeducational services. *Journal of Educational and Psychological Consultation, 22*, 1–20.

Heath, M.A. & Cole, B.V. (2012). Identifying complicated grief reactions in children. In S.E. Brock & S.R. Jimerson (Eds.) *Best practices in school crisis prevention and intervention* (pp. 649–670; 2nd ed.). Bethesda, MD: National Association of School Psychologists.

Jimerson, S.R., Brown, J.A., & Stewart, K.T. (2012b). Sudden and unexpected student death: Preparing for and responding to the unpredictable. In S.E. Brock & S.R. Jimerson (Eds.) *Best practices in school crisis prevention and intervention* (pp. 469–483; 2nd ed.). Bethesda, MD: National association of School Psychologists.

Jimerson, S.R., Stein, R., & Rime, J. (2012a). Developmental considerations regarding psychological trauma and grief. In S.E. Brock & S.R. Jimerson (Eds.) *Best practices in school crisis prevention and intervention* (pp. 377–399; 2nd ed.). Bethesda, MD: National Association of School Psychologists.

Meyers, A.B., Meyers, J., Graybill,, E.C., Proctor, S.L., & Huddleston, L. (2012). Ecological approaches to organizational consultation and systems change in educational settings. *Journal of Educational and Psychological Consultation, 22,* 106–124.

National Center for Education Statistics. (2009). Indicator 1: Violent deaths at school and away from school. *Indicators of school crime and safety: 2009.*

National Center for Health Statistics. (2016). Deaths: Final data for 2014. *National vital statistics reports, 65*(4).

Ortiz, S.O., & Voutsinas, M. (2012). Cultural considerations in crisis intervention. In S.E. Brock & S.R. Jimerson (Eds.) *Best practices in school crisis prevention and intervention* (pp. 337–357; 2nd ed.). Bethesda, MD: National Association of School Psychologists.

Poland, S., Samuel-Barrett, C., & Waguespack, A. (2014). Best practices for responding to death in the school community. In P.L. Harrison & A. Thomas (Eds.) *Best practices in school psychology: Systems-level services* (pp. 302–320). Bethesda, MD: National Association of School Psychologists.

Rossen, E. & Cowan, K. (2013). The role of schools in supporting traumatized students. *Principal's Research Review, 8*(6), 1–8.

Schonfeld, D.J., & Quackenbush, M. (2010). *The grieving student: A teacher's guide.* Baltimore, MD: Brookes.

Worden, J.W. (2009). *Grief counseling and grief therapy: A handbook for the mental health practitioner.* (4th ed.). New York: Springer.

3 Response to Sudden Deaths
A Primary Role for School Psychologists

Anthony Pantaleno

INTRODUCTION

As I look back on my graduate training from 1975–1986, I am grateful for many experiences and for those priceless mentors who taught me everything I needed to know to make school psychology more than a job—a career with the capacity to effect meaningful change in the lives of children. There was one arena, however, where I felt that I was flying by the seat of my pants for many years—a skillset sorely needed and almost completely ignored in the school psychology curriculum of the day—the art of crisis prevention, intervention, and postvention.

DESCRIPTION OF RESPONSES

First Student Death a Homicide

Along with the feelings experienced by the loss of several colleagues over the years, my first close-up experience with student death occurred on Mother's Day in May 1996, approximately 18 years into my career. A public murder took place—a 16 year old student was strangled on the front lawn of her home. I recall the press literally camping out across the street and around the perimeter of the school for months. Many long hours were spent with the closest friends of this young woman.

At that time, our superintendent called in a crisis consultant from the local Board of Cooperative Educational Services (BOCES), who met with our team each morning and afternoon to check our own mental status due to the traumatic and proximal nature of this event. My mother was very ill at the time and near death. The healer became the person in need of healing.

Eight years, and two trials later, forensic evidence finally convicted the student's neighbor. All that time, a dull heartache and an atmosphere of fear hung like a dark cloud above the community.

Grieving a Former Student's Death by Suicide

The next time a profound school crisis knocked on my door was not regarding an active student, but one with whom I had worked closely for four years. During our time together, JJ (initials changed) was one of the most talented Natural Helpers (a national peer helping program) I had ever trained. His ability to listen reflectively to a peer in crisis was unparalleled. He often spoke of wanting to become a psychologist, and this was his primary goal as he headed off to college. During his college years, he developed symptoms of depression, which eventually led him to attempt suicide. After a rescue and treatment at home, his life and career seemed to soar. At age 22, for reasons unknown to those closest to him, he finally died by suicide in January, 1997.

I remember the call from my principal early one Saturday morning as I sat at my computer. "Are you sitting down?" she said. "JJ was found in Central Park. He hung himself". The world stopped. My breath was gone. In that moment, I felt like I would never laugh again. I had no recollection of the rest of the conversation, but only of dropping my head and bursting into tears. The days, weeks, and months that followed created in me a sense of loss, deep grief, personal failure, and doubting belief in my own clinical skills. The years that followed have kept these memories as alive and fresh as if they occurred yesterday.

It was not until close to my retirement in June 2015 that I gained the perspective, time, and energy to do something about helping colleagues who followed me to be better prepared to manage school-related crises. The momentum for this decision was also driven by events of the last three years of my career, especially the 2014–2015 school year, when my small suburban high school of just over 800 students in Long Island, New York experienced the traumatic death of three students under vastly different circumstances.

Death of a Beloved Teacher: Searching for Effective Response

On a cold morning in December 2011, I made the normal 45-minute drive to work. As I entered my school building around 7:15 and approached the lobby, a group of administrators and counseling staff stood in their jackets, more than the usual one or two on a typical morning. I recall thinking it odd that a field trip would require so many chaperones. Approaching my principal to inquire about all the activity, I was stopped in my tracks, "Don't take your coat off. Drop your bag in the office and come with us. One of our teachers died early this morning at home and we're all headed over to the building to set up a crisis response team." What I would learn within minutes is that the teacher was a beloved member of this community, a graduate of our high school, a mother of two sweet little girls, and, as I would always remember her—the softly-spoken and gentle peer helper whom I had trained as a Natural Helper only years before. She apparently died at home in early morning, while getting ready for school. Her second grade class was about to have a very traumatic day, not to mention that those responding were friends and colleagues stricken by our own grief.

What to do first? Teachers were on the way into the building, some knowing and some not knowing. Who would greet the class and guide them through the morning? How many teams would we need to speak to children, colleagues, and parents? Somehow, we came through these events well by all standards. Our loss was deeply felt by all, and perhaps the community knew implicitly that we had done our best to cover all bases on that fateful morning. Our team was exhausted, but pulled together at the end of the day to plan for the coming days and weeks.

It was in these trenches that I was again reminded how my colleagues, administrators, students and parents looked to the school psychologist for leadership during these dark days. Yes, we had our emergency plans like other schools, but one first had to find the plan, then brush the dust from the cover, then dive in and hope for the best. The notion of crisis drills was executing mandatory fire drills or the occasional lockdown—the same things I remember doing as a student in the New York City public schools. Although the global nature of crisis had changed drastically and cases of school violence captured national headlines, schools had not progressed much in the seventies and eighties in developing a more sophisticated game plan for the safety and welfare of students, teachers, and school culture. The function of school psychologists was often limited to evaluating students for placement in special education. We were needed in a central role as the experts in child development, stages of grief and recovery, and our unique connection to staff and families.

The year after the death of our second grade teacher was spent wondering how I might make a difference for my colleagues in future crises. I began to learn about the National Association of School Psychologists' (NASP) PREPaRE curriculum (Brock et al., 2009) and started planning to collaborate with my local BOCES to sponsor the first course. Those interested in training can visit www.nasponline.org/professional-development/prepare-training-curriculum.

Devastating Series of Sudden Student Deaths

My 38th year as a school psychologist started in September 2015—the referral list for initial evaluations, re-evaluations, the growing list of students in need of counseling, the IEP team schedule, and the age-old dilemma from last June—where had I stored my stopwatch? On a late night in October the familiarity of these routines was burst by a phone call from my principal. At a home football game earlier that day, a player from the visiting team had sustained an injury and later died. Tom was a sophomore who was loved in his community. Our building team convened at 6:00 a.m. the next morning before classes to plan a staff discussion on how we would respond in the coming days, as if we had lost one of our own.

As the principal put the telephone and email emergency notification procedures into effect, he said he was looking to me for support on how to manage this crisis. Fortunately, I was good friends with the school psychologist whose district was sharing the mourning process with us, and we began to collaborate as the days moved forward. A number of things raced across my mind. I needed all of my energy to allow these thoughts to come for many hours that night, and do my best to keep focused on a set of strategies that would best serve us.

A much anticipated football game, many spectators witnessed the tragedy: the football players and cheerleaders of both teams, the coaching staffs, the friends, teachers, and families of the players, and members of both communities who just came out to support their teams. Which players from each team were involved in the play that led to the injury? Was the deceased student's family at the field and did he have any school-aged siblings? How would the district convey its condolences in a manner appropriate to model for so many young people?

In the next few days, events following the death of a young person began to unfold. The memorial service and funeral were planned, the family shared requests for a memorial, and teachers shared feelings with their students while attempting to get back to normal. The system attempted to reset, in an environment ripe with the vulnerability that comes from being reminded of the fragility of life.

Not long after these events began to settle, a winter snowstorm engulfed Long Island. This storm became the prelude for our community's next traumatic loss. A few of our students went snow tubing during the late night of the storm, climbing a steep hill rising from a street lamp below. As they began tubing down the hill, one of the boys lost control and went head first into the base of the street lamp. He sustained a severe head injury. His friends brought him for help as fast as they could, but he died as a result of his injury. The student was one of the most well-known and beloved boys in his senior class.

It is during midterm week and classes would not resume for a few days. Again, our building team is called upon to have a crisis plan in place. This time, we decide to invite the students closest to the young man into an open meeting at school before classes resume. I recall sitting around a long table in the library with some of Sean's closest friends, including two of the boys who were snow tubing with him, his school counselor, and some members of my crisis team. I recall the colorless faces that had already been crying this out for days. I brought some blank paper and crayons, and invited the students to do whatever they wanted with them. After some time the strain of the day settled, and the first few students took up the challenge to express themselves, with others following. What they produced was not as important as the chance to have their voices heard—to say something about the friend they had lost, to have some closure.

I followed Sean's schedule on the day classes resumed, to offer condolences and to offer a message of hope. It was the most difficult challenge of my career. The teachers were unable to say much, their faces ashen, a lifeless silence hanging in the air. I read to them one of my favorite short stories by Leo Buscaglia, entitled *The Fall of Freddie the Leaf* (1982). I did not ask for reactions, only for the students to think about what was most precious to them. During this time, other crisis team members were conducting triage, psychological first aid, and other interventions for students and staff.*********⸺⸺⸺

Teachers Share Their Perspectives of a Tragedy

At the end of the school year, I asked some of Sean's teachers to try and recount their personal recollections. Here are two anonymous, slightly edited replies. These convey the heart and soul of what we face with the reality of trauma:

Teacher #1:

My first thought when I heard Sean's death was complete shock. Oddly, I turned off my phone the night before. My husband was traveling with the storm. He arrived back on the island as the roads were closing.

In the morning, I woke up to my fellow teacher and best friend calling my husband. It was around 7:30 a.m. and she knew that my phone was off. As soon as I turned on my phone, I saw a thread of about 25 messages from various teachers, my friends. They were looking for me and knew that I was going to be upset because I was very close with Sean.

My heart sunk. I was in disbelief. I checked News 12 to see if it were true. Before I spoke to any of my teacher group, I called Sean's school counselor. I knew she would know. When she confirmed it, I sat in shock. I could not believe such a nice boy had died.

I then reached out to the athletes on my team as their coach. They were very close with him; one of the girls was dating him. They were a mess.

My son was nine months old so I went to make him breakfast. I thought I could be normal but I wasn't. I looked at my innocent son and I lost it. I cried for hours. Quietly, my husband took out Sean's papers from my bag. I had tons of grading and his life studies paper was in my bag.

Sean was obsessed with the afterlife. Our final conversation was about how I should believe in it. I did not say how I really felt, but he was trying to convince me to believe. That was on January 22.

As the day went on, group messages went back and forth in disbelief. The male teachers were not talking much. At around 6:00 p.m., a good friend and colleague who does not discuss his feelings often, called me to see if I was okay. It was evident he wasn't. We both tried to digest it. We discussed changing seats so the kids (and us) did not have to look at Sean's empty seat.

Sean's death deeply impacted me. I felt responsible for the seniors on my team, in my classes and in Sean's class. When we went back to school, just the teachers, I sat in my room moving desks back and forth. I took so much time trying to figure out a way to help my class help me.

I changed posters in my room. I removed the lame teacher posts and put up all quotes about life and love written by my favorite authors.

I knew that the kids could want the desks back. What we did with the desks was going to be completely up to them. I now know that they appreciated this. I also gave each one a bright card. If they did not want to get called on, I told them to put the card on their desk. I told them that they could use this technique for the rest of the year.

I am going to be honest. I do not feel that I was adequately prepared for what we had to do that day and the weeks that followed. I think the fact that Sean's funeral was before we actually returned to school helped the kids and the staff in a way. The kids already saw us at the wake and the funeral. Now at school, I think they wanted to get back to the familiar routines. Because it took a week to see them because of weather and midterms, the kids started to band together. Sean's friends started to ice out the other kids. This divide among the students did not last long, but while it was there, it was awkward.

I think the kids looked to us as their teachers to help them, and they did not go to the mental health professionals. I do not know why. As the year went on, I literally felt drained. I grew very close to this grade, but I needed them to graduate and move on.

I think if the school had more people to come in and work with us in the following months, to show the kids multiple times that they were there for them—that would have been much appreciated. All of the support came in the initial aftermath and then seemed to quickly recede.

I think that I would tell my colleagues to be themselves in front of the kids. Show them how you feel but know that means they will show you how they feel. I will be honest. I felt very close to every one of my students last year. They were kind to me and we share a bond because of Sean.

Teacher #2:

When I heard of Sean's death, I immediately thought of his family and the friends he was with when he died. I remember feeling a sense of shock when I saw it on the news that Sunday morning, and then when I read the email from Sean's school counselor, it seemed so unbelievable. Personally, Sean's death reminds me of "the big picture" in life—what is important is the relationships/connections you make with the people around you.

I knew it was going to be tough when we met again as a class. You might recall that a snow day followed, then we had midterms, so I didn't formally see my students in the classroom until a week later. I did communicate with some of them through email, and I saw a lot of them at the wake/funeral and the memorial we had at school with his family. I figured it would be a day to talk, cry and grieve. I brought tea, hot chocolate, and cookies. My other class was just as distraught as the class Sean was in.

In Sean's class, I rearranged the desks in a big circle. For me, the thought of seeing his empty desk in its usual spot would have been too much. I felt adequately prepared because I was a counselor in a former life, and have helped people through loss and bereavement. I also let myself cry with them. Others might not feel comfortable doing this.

It wasn't very difficult to start teaching; to be honest, the students were ready to focus on school work and they knew they could leave the room if they were having a tough time. The difficulty again, was Sean's empty desk. It was very sad for us all.

I really did feel supported by the school, but I feel it is something you have to deal with in your own way. Each teacher might deal with it differently, and it depends on the dynamics of the class, and at what point something like this occurs during the school year. I meet with some of my students for two periods every day, and I get to know them really well, so I felt comfortable grieving with them and helping them cope. It was very helpful to have you as a resource. Having the principal look in on us periodically, and sit with us as a class also helped.

If my colleagues were to go with something similar, I would advise them to go with their gut instinct, their heart, and seek help from the great resources they have.

*   *   *   *

As the end of the school year approached and May turned the corner, I hoped that we had seen enough tragedy for one year, but that was not the case. A graduate from the previous school year sadly passed at this time, allegedly from a heroin overdose. The impact on the current students was less eventful as the victim was a former graduate and not part of the school culture that had been traumatized. But the classroom teachers who knew the student took the brunt of this loss once again. As a faculty, we had passed through a year in hell. The lessons learned could only help us in the future, but not a single staff member was sad to see the school year end.

INSIGHTS AND ACTION

As a veteran school psychologist, I developed a false sense that I had seen and heard it all. A glimpse of the raw emotions and personal toll that these events took on our entire community could not be gleaned in graduate seminars, professional journals, or any workshop. Not only did I begin to grasp the scope of what was needed in an effective crisis response, but in my heart, I knew that much more was needed, particularly for the caretakers on the front lines in the classroom, those closest to the student scene every single day—the teachers.

Recently, a group of concerned Long Island school psychologists and other crisis team members began to collaborate, forming a regional crisis team modeled after work by the Putnam Northern Westchester BOCES Regional Crisis Team after the Columbine shootings in 1999. Our group sought to standardize the training received by all crisis responders so that all spoke the same language and implementation error was minimized during a school crisis. We also sought a means whereby school districts could band together in a neighbor-to-neighbor effort to assist each other when any one of us experienced an overwhelming event that might stretch our individual resources to the max. This was the beginning of the Long Island School Practitioner Action Network (LISPAN), whose mission statement speaks to the following goals:

1. To provide temporary additional mental health supports and volunteer mental health staff to LISPAN member school districts, upon request of the superintendent or designee, to mitigate the impact of a local or regional school crisis;
2. to improve the communication and clinical response effectiveness of LISPAN volunteers by hosting a monthly meeting of network representatives at rotating member district locations;
3. to facilitate the implementation of a "best practices" model using the National Association of School Psychologists' PREPaRE workshop 2 (Brock, 2011), "Crisis Intervention and Recovery: The Roles of the School-Based Mental Health Professional". All LISPAN member districts agree to have at least one staff member who has completed PREPaRE workshop 2, which will be offered at least annually by one of the regional BOCES;
4. to emphasize the role of crisis prevention via ongoing training in PREPaRE workshop 1 (Reeves et al., 2011), "Crisis Prevention and Preparedness: Comprehensive School Safety Planning";
5. to serve as a model for the formation and promotion of regional crisis teams.

The membership documents and resources assembled by LISPAN may be viewed at its web site, www.lispan.org.

Myths of School Crisis Prevention

For your consideration, I offer three myths of school crisis prevention that I believe are widely held in our current school culture. It is towards the elimination of these myths that LISPAN and other school-based regional crisis teams find their greatest challenge.

1. It can't happen here, or what are the odds . . . ?

With the employment of School Resource Officers, the installation of metal detectors, and the investment made in digital technology, there is a false sense of security in many schools that "all is well"—that those one or two (unarmed) folks standing in the lobby behind a podium with a walkie-talkie can field just about anything. How long will we endorse this fantasy?

2. We can do this alone.

Until recently, schools were wary to bring in "outside experts" to help manage a major crisis. There was a huge lack of trust, not only of unknown professionals coming in, but in the community perception that backup somehow meant the in-house health team would be perceived as incompetent. Some administrators also felt a loss of control handing over the reins to others, no matter how credible their expertise. However, realizing the vulnerability of our schools and the need for reaching beyond our borders in time of need, we must come to terms with a harsh reality. We cannot do this alone.

Zenyep Tufecki (2015), writing in the New York Times, states "After studying 160 'active shooter' events over the past decade, Andre Simons, of the F.B.I.'s Behavioral Analysis Unit, concluded that "'the copycat phenomenon is real . . . We think we're seeing more compromised, marginalized individuals who are seeking inspiration from those past attacks.'"

3. School mental health professionals have already learned how to respond to school crisis events in graduate school. Why get them involved in yet another training curriculum when they already have so much else to do?

Most school psychology and counseling curricula across the country do not contain dedicated courses in crisis management. School psychologists and counselors take it upon themselves to find this training after graduation, or do not seek any formal training at all. With many other responsibilities, a typical practitioner might have to dust off a crisis procedure manual just as a major crisis is unfolding. Until the feelings of managing a live crisis situation have been part of our learning and we see the need more clearly, we may take the easier path, choosing to leave school crisis prevention work to other first responders or to our administrators.

School Crisis Response Planning

Due to the importance of school emergency preparedness, the New York State School Safety Improvement Team recommended sweeping statutory amendments to broaden the scope of crisis planning. Highlights of these regulations include:

1. All staff are required to have annual school safety training on the district emergency plan, including components on school violence and mental health.
2. District-wide safety plans are required, including designation of a chief emergency officer responsible for coordinating communication between school staff, law enforcement, and first responders.
3. Building-level emergency plans must include procedures for response to situations requiring evacuation, sheltering and lockdown (evacuation routes, shelter sites, medical needs, notification of parents/guardians). The building-level response team is expanded to include fire officials, and at school board discretion, a student may participate on the school safety team with limitations involving confidential information.

LESSONS LEARNED AND REFLECTIONS

As the country and the world have witnessed the violent transformation of a peaceful scene into one of chaos, pain, and human suffering, it behooves all state departments of education to consider whether efforts to safeguard the sanctity of educational institutions, particularly those serving children, are being enacted. School psychologists become a critical part of the fabric of this response, perhaps not by our intention when we set foot into the field, but as the result of events demanding that our roles and functions change and expand. No one ever asked this writer in 1978, my first year in the field, if I could envision myself providing leadership in the crises described in this narrative. It became a necessity. It became a calling. It became a meaningful value upon which our core professional ethics as school psychologists had been founded.

As this new age of school crisis readiness dawns, I find that music serves as a powerful balm on days when we feel the weight of what we do. I am reminded of a song lyric recorded by the Beatles on August 18, 1969 at the conclusion of *Abbey Road*: "And in the end, the love you take is equal to the love you make". So what will be our legacy as school psychology practitioners and crisis team members? Will we find the love in our hearts to create the safety nets that our children, teachers, parents, and communities deserve, to respond decisively and efficaciously to an unfolding school crisis? Will we be able to tackle the many challenges that this vital aspect of our daily work asks of us? If your school is like mine, funding for professional development is not what it was 20 years ago, and time out of the building now requires signoff from multiple administrators. So what lies before us is not an easy path, but neither is the decision to stay on the familiar path, only to receive that phone call which will teach us the lessons we need to learn the hard way.

I close this piece with some personal reflections from my career in the trenches:

1. I am the best advocate for school crisis prevention education. Use my connections wisely, networking with colleagues. Look to the work done before me. No need to reinvent the wheel.
2. I am emotionally vulnerable if I try to respond to crisis at the same time other personal events are impacting me; accept this reality and share it with my team. Accept whatever support I need during a crisis as freely as I give my support to others.
3. While I was not taught in graduate school how to manage *colleagues* who are in crisis, I now know that this is a first priority of school psychologists. Since our fellow educators will be the ones to whom students and parents will turn, give care to the caregivers, tell them your door is always open, and take time to visit their classes regularly to let them know they matter. Continue to do this until they cue you that support is no longer needed.
4. I am grateful for taking the time to develop a daily mindfulness practice, and for my personal spiritual beliefs. These are rocks during the storm.
5. I will not know all the answers in responding to a particular crisis event, but someone will. Be humble and ask.
6. Take frequent breaks as the situation allows to eat, drink, use the rest room, get some fresh air.
7. As much as possible, when I go home at the end of the day, do not bring the events of the day home with me.
8. Allow music, art and any of my other favorite pastimes to find some space in my day.
9. Remember that the pain of this time will slowly pass and that people are basically good.
10. Above all else, in times of darkness, find some humor when appropriate and where you can. The best intervention we have is our smile and our presence.
11. Encourage early career school psychologists to seek PREPaRE training and to encourage their peers to do the same.
12. Please take some time to read the most detailed, heart-wrenching story of the events in Newtown as chronicled in Matthew Lysiak's *Newtown: An American Tragedy* (2013). Every one of us will relate to the unfolding events of this story and will ask many times along the way, "What would I have done"?

Crisis prevention will continue to be a key role for all school psychologists, whether we seek it or not.

This personal narrative is dedicated to the beloved memory of our dear Sandy Hook colleague, School Psychologist Mary Sherlach.

REFERENCES

Brock, S.E. (2011). PREPaRE Workshop 2: Crisis intervention and recovery: The roles of school-based mental health professionals (2nd edition). Bethesda, MD: National Association of School Psychologists.

Brock, S.E., Nickerson, A.B., Reeves, M.A., Jimerson, S.R., Lieberman, R.A., & Feinberg, T.A. (2009). *School crisis and prevention: The PREPaRE model.* Bethesda, MD: NASP.

Buscaglia, L. (1982). *The Fall of Freddie the Leaf.* Thorofare, NJ. Slack Inc.

Lysiak, M. (2015). *Newtown: An American tragedy.* New York: Gallery Books.

Reeves, M.A., Nickerson, A.B., Conolly-Wilson, C., Lazzaro, B., Susan, M.K., Pesce, R.C., & Jimerson, S.R. (2011). *PREPaRE Workshop 1: Crisis prevention and preparedness: Comprehensive school safety planning.* (2nd ed.). Bethesda, MD. National Association of School Psychologists.

Tufecki, Z. (August 27, 2015). The Virginia Shooter Wanted Fame. Let's Not Give it to Him. *The New York Times:* retrieved from www.nytimes.com/2015/08/27/opinion/the-virginia-shooter-wanted-fame-lets-not-give-it-to-him.html?_r=0

4 Death in the Classroom
Severe Emotional Trauma

Jeffrey C. Roth[1]

INTRODUCTION

While I didn't want to lose the day in my schools, something told me I should attend the workshop at a firehouse in southern Delaware titled, "Hearts Behind the Badge: First State CISM Team". I learned about Jeffrey Mitchell's Critical Incident Stress Management (CISM) model and how it could help mitigate stress reactions among traumatized first responders (Everly & Mitchell, 2003, 2010; Mitchell, 1993; Mitchell & Everly, 1998).

Severely Traumatic Incident

During the presentation, I heard my beeper as we did not have cell phones at that time. It indicated 911 and the phone number of our lead secretary. I left the presentation to call the secretary and heard that a student collapsed and died in one of our middle schools. I could feel the color drain from my face. I returned to the workshop to say quick goodbyes and ran to my car. I remember nothing of the drive upstate to the scene of the sudden death.

PREVENTION AND PREPAREDNESS

Traumatic events are often sudden and cannot be prevented. Recovery after an unexpected tragedy frequently depends on a combination of factors such as the school climate and the extent to which students feel cared for and have a sense of belonging. The extent to which students feel they have a role in the recovery process is also relevant (PREPaRE Model: Developing Resilience and School Connectedness). Jimerson and colleagues (2012) point out that while prevention may not be possible, a comprehensive crisis plan that is flexible and includes school and district crisis teams prepared for a wide range of unusual events such as sudden death on campus can mitigate emotional trauma, support coping, and educate caregivers.

Reeves and colleagues (2012) suggest that a key element of crisis preparedness involves conducting a *vulnerability assessment* to determine potential environmental risks. This assessment reviews preventive measures and potential hazards inside and outside of the school building, strengths and needs relating to the students' physical and psychological safety, and review of the school's emergency communication procedures. Following the assessment, planning can encompass universal prevention programs and targeted, incident-specific response protocols (Reeves et al., 2010; Reeves et al., 2011) (PREPaRE Model: School Building Vulnerability Assessment).

After crisis teams are trained, the teachers and support staff can be taught constructive ways to respond to grief reactions with developmentally and culturally appropriate interventions (PREPaRE Model: Psychological Education: Caregiver Training). Systems can be developed to assess and address student needs during the response (PREPaRE Model: Evaluating Psychological Trauma). This includes teaching parents to help their children cope, and coordinating with community agencies and professionals. Crisis response planners can flexibly evaluate the varying effects of sudden death on students and staff, and can include long term follow-up (Jellinek & Okoli, 2012; Jimerson et al., 2012).

Preparedness after a vulnerability assessment often involves planning and implementing drills or exercises (PREPaRE Model: Exercising School Crisis Plans). When drills are properly conceived and practiced, they enhance student and staff knowledge and skills about response to emergencies, without increasing anxiety (Zhe & Nickerson, 2007). Unexpected, less frequent events such as a sudden death on campus, may not lend themselves to drills which require active student involvement. These types of severely traumatic incidents can be practiced through discussion-based exercises such as tabletop drills and workshops (Reeves et al., 2010).

Our district crisis team had been gaining experience with higher incidence events such as the off-campus death of students, teachers, and parents, but never the sudden, incomprehensible death of a student in a classroom in the presence of his peers. We had not conceived, drilled or practiced response to such a severely traumatic incident. This was prior to training in the PREPaRE model (Brock et al., 2009). There were no protocols in place. Our inexperience and lack of practice led to some mistakes during the response, and lessons learned.

DESCRIPTION OF RESPONSE

As I was driving back from the workshop, members of the district crisis team joined the school team providing psychological first aid to students. School was being dismissed by the time I arrived. Here's what happened. Shortly after lunch, the thirteen year old had gone to his math class and remarked to the student in front of him, "I feel funny." He collapsed. His classmates' laughter at his presumed prank turned into horror when they realized he had lapsed into unconsciousness. Hearing the commotion, eighth grade teachers rushed to join the shocked math teacher and her students. The school nurse was quickly summoned and CPR performed, but the student was not revived.

The principal, nurse, and district responder Marty Tracy joined the family at the hospital, where the boy was pronounced dead. Weeks later, we learned that the death resulted from a previously undetected heart defect. In their grief, the family reached out and kept the principal informed of funeral arrangements.

The boy was popular, a likeable student with an endearing manner and ready sense of humor. His death was grieved throughout the school. For our team, this was a uniquely challenging experience due to the severity of emotional trauma and exposure to student witnesses, their teacher, and the first responders—other teachers, counselors, the school nurse and administrators. *Primary triage*, which identified student and staff witnesses, was relatively obvious, but the task of working with these traumatized individuals was exceedingly difficult (PREPaRE Model: Primary Triage: Evaluating Psychological Trauma).

Jimerson and colleagues (2012) observe that following a sudden death there is greater risk for psychological trauma among those who witnessed the event, relatives or friends of the deceased, and those with early developmental levels, mental illness, or limited social support (Brock & Davis, 2008). Brymer and colleagues (2012) suggest that psychological first aid providers prepare to support "individuals whose reactions are so intense and persistent that they affect the students' ability to function or be oriented to their current environment." Strategies such as minimizing exposure to the traumatic incident (PREPaRE Model: Emergency Procedures that Minimize Exposure), simple grounding techniques, breathing relaxation exercises, or cognitive restructuring can help calm students who witness a devastating event (Brymer et al., 2012).

Crisis Counseling, Team Debriefing, Faculty Meeting

Crisis counseling began immediately in the school library and in smaller adjacent rooms with student witnesses (PREPaRE Model: Provide Interventions and Respond to Psychological Needs). The blended school and district teams provided psychological first aid, identifying students appearing most affected and needing further intervention. Counseling was also available for affected teachers, but the focus was clearly on the students.

After school, we debriefed as a blended team, joining the school administration and counseling staff to plan our response. The principal composed a letter to the entire school community and a separate letter for the parents and guardians of the students in the affected classroom. School counselor Leslie Carlson proved especially able and assumed a leadership role. She later joined the district team, eventually becoming one of its coordinators.

Before leaving on the first day, we triaged further, reviewing the names of all students who had been counseled, discussing those who appeared to need small group or individual follow-up the next day. We divided their names among crisis team members for parent/guardian telephone contact (PREPaRE Model: Secondary Triage: Evaluating Psychological Trauma).

We held a faculty meeting—with coffee—the next morning and shared what resources were available for student and staff counseling, how to support students and each other, and when and how to refer students needing more attention (PREPaRE Model: Caregiver Training). The superintendent provided floating substitutes, available if teachers needed relief. We distributed to staff information about grief reactions and a handout on talking with students about death and grief. We advised about escorting distressed students to the library, where crisis counseling was available. Teachers were reminded to monitor students not only for the most obvious grief reactions such as sadness and crying, but also for other behaviors such as being unusually quiet, withdrawn, angry, or acting out. Plans were made for school counselors familiar to the students, reinforced by district responders, to follow the deceased youngster's schedule to support students and teachers.

After the faculty meeting, the team stationed members at various strategic locations such as buses, hallways, faculty lounge, and library to escort students needing continued crisis counseling. Some counselors were assigned to facilitate small groups while others met with individuals. Everyone continued documenting names of students seen and their emotional status. Responders were assigned to the affected teaching team's corridor in the event that extra support was spontaneously needed.

Each class offered opportunities for structured discussion and for monitoring the emotional status of students and staff who were witnesses. The PREPaRE model (Brock et al., 2009) offers structured group and individual interventions, including some especially useful for an affected classroom. "Student Psychoeducational Groups" provide an opportunity to answer questions and dispel rumors before focusing on instruction about common crisis reactions and plans for stress management. "Classroom-Based Crisis Intervention" (CCI), offers the option of a longer, more intensive session for students exposed to the same traumatic event. It is recommended that participation be voluntary and by parental consent. CCI provides an opportunity for similarly affected students (homogeneous groups) to share their stories, identify and normalize common reactions, and identify maladaptive reactions requiring follow-up. A key of CCI is to empower participants with coping strategies, information and individualized plans for stress management, and to follow-up with parents and caregivers with information about how they can support coping. Finally, for distressed students in need of individual attention, PREPaRE offers a structured "Individual Crisis Intervention" (ICI), which reestablishes coping, provides support, identifies crisis-generated problems, and begins adaptive coping and problem solving. ICI affords an opportunity to assess the student's level of risk for stress reactions or potential self-harm.

Throughout the first two days after the death, students had an opportunity to express their grief. They shared thoughts and feelings about the incident and how it was affecting them. They were exposed to many caring adults who actively listened, empathized, and taught about the grief process and stress management strategies. They were encouraged to talk with their families, who continued to be contacted with information about their children's status and ways to support them (PREPaRE Model: Reestablish Social Support Systems/ Caregiver Training). Students appearing most at risk for psychological trauma were referred for outside treatment.

Eventually, students receiving crisis counseling began sharing pleasant memories of their classmate, and began writing anecdotes and drawing pictures for inclusion in a memory book to be presented to his family. As in previous crises, the art teacher provided plenty of materials. There was also a paper mural taped to the wall in an accessible, easily monitored hall location where students could express thoughts in prose, poetry and drawings. Murals created by the students were given to the grieving family. Creative expression helped students feel a sense of control, working individually or in groups (PREPaRE Model: Providing Opportunities to Take Action). Gradually, they returned to their regular class schedule. More and more, normal routine was reestablished.

Media Boundaries: Intrusion

A number of incidents were unique to this tragedy. We awaited word about the funeral while diligently seeking a return to normal routine. While I monitored the cafeteria during eighth grade lunch, an upsetting incident involved a news photographer accompanied by the district superintendent. The superintendent had been well meaning and supportive, except on this occasion. The *News Journal* photographer snapped photos of students as they ate lunch. The scene took on a surreal quality, an intrusion far from normal routine.

I felt anger and confusion approaching the superintendent. I said, "What is this newspaper photographer doing here?" His response in the noisy cafeteria was a quizzical look and a shrug. I felt powerless to pursue the matter with him. As the district authority, he had his reasons. I momentarily lost my ability to reason, because I proceeded to approach and bargain with the photo journalist. "I wish you weren't in here doing this," I said to him.

"What's the problem?" he asked. "We are trying to establish a normal routine for these students and you are in here taking photos," I explained.

I thought surely he realizes it is not normal routine for his large, obtrusive camera and tripod to be reminding these kids about what happened while they are at lunch. Maybe I was being too sensitive. I wanted to leave the cafeteria.

It felt so intrusive, so lacking in respect. If there was an agreement with the media, trading photos to establish broader boundaries, I did not know about it and gave up speculating why it was happening (PREPaRE Model: Media Relations Protocols/Collaborating with the Media).

Funeral Plans

As the family liaison, the principal received information about the funeral. Our position was to respect the wishes of the family regarding whether the service would be private or open to the public, including students. The family decided to welcome the public. The district provided substitutes for teachers attending the funeral. We sent letters home with details and asked that responsible adults accompany students wishing to attend and stay with them for the rest of the day rather than return them to school. Before the funeral, we offered counseling for students attending the funeral to help them prepare and address cultural and religious differences they might experience.

School and district staff would disperse themselves throughout the crowd at the church, to be in proximity to students. We anticipated that some students would attend without parents or guardians. On the morning of the funeral, we got word that a well-meaning bus driver had offered to pick up students at bus stops and deliver them to the church. He was told not to proceed, that his kind offer was counter to the school's wish that responsible family members accompany the young mourners. In response to concerns that students might be waiting at the bus stops, team member Elliot Davis and I circled to every stop several times, checking for students. None were waiting. We drove past the church where a crowd filled the steps and spilled onto the entire block. I learned from those who attended that the service seemed to provide comfort for the family, students, and staff. There would be an opportunity for students to talk about the funeral and their feelings at school the next day.

RECOVERY

Memorial Versus Return to Normalcy

An incident unique to this crisis provided a source of learning. Students from a classroom near the one where the student collapsed wanted to have a large memorial in the gymnasium soon after the funeral. During debriefing, the crisis team discussed the situation with the school administration and reached consensus that the nature and timing of the proposed memorial might interrupt the emotional healing that began at the funeral, and might interfere with the timely return to normal routine. While not all students attended the funeral, the service did signal a degree of closure. Individual and small group counseling, on or off campus, continued for those "at risk", but the team feared that a large memorial ceremony after the funeral might cause the school's grieving process to regress. The team decided that a memorial toward the end of the school year could be planned with student input. Though it was a difficult decision, we believed it was best from our perspective at the time (PREP<u>a</u>RE Model: Special Considerations When Memorializing an Incident).

I walked along the main corridor, beginning to feel a sense of relief with disengagement imminent. I was approached by an eighth grader walking toward me quickly, purposefully, with a serious demeanor. A few feet away she said, "Are you a leader of the response team?" "Yes," I smiled weakly, preparing to go into counseling mode. Her voice had a mixture of assertiveness, anger and tremor. I sensed she was near tears. "You should let us make a memorial for our classmate. He was our friend. This is *our* school."

I was taken by surprise. My immediate reaction was to be as non-defensive as possible, realizing that her strong feelings needed to be respected, even though we did not agree. It took courage for her to approach an authority figure and stranger.

"I appreciate that you are telling me about your feelings. That is not easy and takes courage. We are not saying there will be no memorial. We just feel the timing might not be good right after the funeral. There will be a memorial planned for later in the school year." She walked away, clearly dissatisfied. I wondered if we were doing the right thing.

Stating guidelines for delivering psychological first aid in schools, Brymer and colleagues (2012) emphasize the importance of listening carefully when students, parents and staff want to talk, and understanding what they tell you and how you can help. At that point I focused on listening, and then went to the team.

A fellow coordinator, social worker Kitty Rehrig, had also been confronted in protest. It appeared that one of the eighth grade teachers resented the actions of the crisis team and was apparently expressing her anger toward us through her students. Perhaps she felt we were overstepping boundaries by controlling a process that should have been left to the teachers and students. Perhaps her resentment was legitimate. We tried to keep teachers informed, but did not always seek teacher or student input for decisions that directly affected them. I recalled the "free floating" anger that often accompanies traumatic incidents. Whether the teacher was expressing a desire for more control, or other emotional needs, her anger was real and directed toward the team. Initially, we were annoyed that she was "stirring up her students". If she had complaints, why not come to us directly?

While we resented the teacher's apparent efforts to sabotage our good work, we needed to get over our own feelings and try to understand what was happening (PREP<u>a</u>RE Model: Examine Effectiveness of Crisis Prevention and Intervention). Problem solving mode was needed. We were there to do psychological first aid, not psychotherapy, but as crisis responders dealing with emotional trauma, we sometimes needed to analyze reactions in therapeutic terms. When clients are expressing anger, fear, or powerlessness toward us, we need to recognize our feelings, perhaps debrief, and then explore our clients' feelings. Their feelings might provide a clue that helps us figure out and address their needs. Brymer and colleagues (2012) point out the importance of "practical assistance", including problem solving to identify and clarify immediate needs, and to develop and implement an action plan to address those needs. There are times when the crisis team must clarify and solve problems arising out of its own response in order to help those affected.

An immediate issue we needed to explore was finding ways to empower the students and teachers. Providing them with information was important, but we also needed to seek more input when decisions affected them. The school administrator was in the best position to make some decisions, with consultative input from the blended crisis team. Students and staff could provide input or decide memorial plans within parameters set by the building administration. A faculty-student advisory committee could plan for a memorial. While this was a good solution to the immediate problem, another level of exploration and understanding remained necessary.

Was the teacher's anger and need for control also a cry for help—an expression of traumatic stress? Was she fixated on anger that a young student died in class? Was she replaying the powerlessness of running into a classroom where he lay unconscious and could not be revived? The students who witnessed the incident were receiving lots of care and counseling. The adults, especially the teaching team and first responders had received far less support. They appeared at risk for post-traumatic stress reactions. We needed to offer the angry teacher and the other adult responders more support.

A Valuable Resource

I thought back to the workshop downstate for police and firefighters on Critical Incident Stress Management (CISM) and the 911 call on my pager. The workshop I had reluctantly attended might now provide a valuable resource to address the suffering school staff. I had learned that the CISM team was on call to travel regionally to help those at risk for stress reactions. They were trained to provide Critical Incident Stress Debriefing (CISD) which attempts to mitigate the symptoms of post-traumatic stress reactions. I spoke with the school administrators, school nurse and counseling staff about the CISM team and the work they do. Their reaction was quick and unanimous. They asked that we invite them to offer a voluntary session with the teaching team, school administrators, counselors, school psychologist and nurse—about fourteen people. Nearly all would participate, including the "angry" teacher. The session was limited to staff who were on the scene of the incident, the witnesses, the first responders, the most vulnerable.

I contacted the leader of the CISM team and she responded as if she'd been waiting for my call. "You looked upset when you left the workshop. We worried about you." Her welcoming response made it easy to set a date and time. The CISM team of four trained individuals arrived shortly after the end of school on the appointed day. I had set up a quiet space in a room adjoining the library and arranged a circle of comfortable chairs and a table with water and healthy snacks, compliments of food service. Boxes of tissues were available. I welcomed the CISM team and left the school.

I was told that the work done that afternoon was helpful for those involved (PREPaRE Model: Caring for the Caregiver). Several days later, I was told by one of the two school secretaries that they resented not being included in the session. They had been very involved in the event, had made and received crucial phone calls. Should they have been included in this "homogeneous" group of witnesses and first responders? Was it an oversight not to include them? Perhaps another mistake and "lesson learned." Perhaps not. . .

LESSONS LEARNED AND REFLECTIONS

1. Attend workshops to enhance your knowledge and to network with others. What you learn may have an immediate positive impact on your morale and practice. *How might you develop a professional development plan that addresses personal and district needs?*
2. Sudden, severely traumatic incidents can place unique, extraordinary challenges on a crisis team and school. *In what ways can a team prepare for rare, catastrophic incidents, as well as the difficult, more frequently confronted events?*
3. Always address the emotional needs of teachers, administrators and staff, who must be stabilized in order to care for the students. *How can triage be planned and conducted to meet the psychological needs of the entire system— students, staff and community?*
4. Staff monitoring students for reactions that warrant referral must be aware of obvious signs of sorrow (i.e. crying), and less obvious behaviors such as being unusually withdrawn, angry, or aggressive. *What are some ways to educate staff to recognize trauma and grief reactions?*
5. Regardless of the most proactive plans for working with the media, expect the unexpected and be prepared to cope with it. *What are some ways to establish and enforce media boundaries?*
6. In developmentally and situationally appropriate ways student input, including service on an advisory committee, may be helpful when planning a memorial event. *What are some ways to get engage students when planning a memorial?*
7. Planning for a funeral, including preparation of students for cultural and religious differences, and requiring responsible adult supervision, does not mean all will go as planned. *How can advance planning address community needs—those attending the funeral, not attending, and those needing adult supervision?*

8. When faculty or students struggle with misdirected anger and other emotions, responders can view such emotions as diagnostic. Responders can devise ways to understand and resolve underlying needs and conflicts. *During a crisis, what are ways that responders can diagnose and plan helpful interventions to address evolving needs?*

9. Consider who may need outside help, including teachers and support staff. *How might the crisis team take inventory and contact community agencies, law enforcement, and crisis management service groups that could be called if needed?*

NOTE

1. Adapted from *School Crisis Response: Reflections of a Team Leader* (Roth, 2015)

REFERENCES

Brock, S.E. & Davis, J. (2008). Best practices in school crisis intervention. In A. Thomas & J. Grimes (Eds.), *Best practices in school psychology* V. Bethesda, MD: National Association of School Psychologists.

Brock, S.E., Nickerson, A.B., Reeves, M.A., Jimerson, S.R., Lieberman, R.A., & Feinberg, T.A. (2009). *School crisis prevention and intervention: The PREPaRE model.* Bethesda, MD: National Association of School Psychologists.

Brymer, M.J., Pynoos, R.S., Vivrette, R.L., & Taylor, M.A. (2012). Providing school crisis interventions. In S.E. Brock & S.R. Jimerson (Eds.) *Best practices in school crisis prevention and intervention* (pp. 317–336; 2nd ed.). Bethesda, MD: National Association of School Psychologists.

Everly, Jr., G.S., & Mitchell, J.T. (2003). *CISM: Individual crisis intervention and peer support* (2nd ed.). International Critical Incident Stress Foundation. Baltimore, MD.

Everly, Jr., G.S., & Mitchell, J.T. (2010). *A primer on critical incident stress management (CISM).* Ellicott City, MD: International Critical Incident Stress Foundation.

Jellinek, M.S. & Okoli, U.D. (2012). When a student dies: Organizing the school's response. *Child and Adolescent Psychiatric Clinics of North America, 21,* 57–67.

Jimerson, S.R., Brown, J.A., & Stewart, K.T. (2012). Sudden and unexpected student death: Preparing for and responding to the unpredictable. In S.E. Brock & S.R. Jimerson (Eds.) *Best practices in school crisis prevention and intervention* (pp. 469–483; 2nd ed.). Bethesda, MD: National Association of School Psychologists.

Mitchell, J.T. (1993). Critical Incident Stress Management: The First Decade. *Life Net, A Publication of the International Critical Incident Stress Foundation, Inc., 4(4),* 1, 2.

Mitchell, J.T., & Everly, G.S. (1998). *Critical incident stress management: The basic course workbook* (2nd ed.). Ellicott City, MD: International Critical Incident Stress Foundation.

Reeves, M.A., Conolly-Wilson, C.N., Pesce, R.C., Lazarro, B.R., & Brock, S.E. (2012). Preparing for the comprehensive school crisis response. In S.E. Brock & S.R. Jimerson (Eds.) *Best practices in school crisis prevention and intervention* (pp. 245–264; 2nd ed.). Bethesda, MD: National Association of School Psychologists.

Reeves, M., Kanan, L., & Plog, A. (2010). *Comprehensive planning for safe learning environments: A school professional's guide to integrating physical and psychological safety— Prevention through recovery.* New York: Routledge.

Reeves, M.A., Nickerson, A.B., Conolly-Wilson, C., Lazzaro, B., Susan, M.K., Pesce, R.C., & Jimerson, S.R. (2011). PREP<u>a</u>RE Workshop # 1: *Crisis prevention and preparedness: Comprehensive school safety planning* (2nd ed.). Bethesda, MD. National Association of School Psychologists.

Roth, J.C. (2015). *School Crisis Response: Reflections of a Team Leader.* Wilmington, DE: Hickory Run Press.

Zhe, E.J. & Nickerson, A.B. (2007). The effects of an intruder crisis drill on children's self-perceptions of anxiety, school safety, and knowledge. *School Psychology Review, 36,* 501–508.

Section 3

School Crisis Response to Death of a Teacher/Staff Member

INTRODUCTION

Some of the most difficult and complicated school crisis responses follow a teacher's death, whether sudden or after a terminal illness. Teachers and staff members often endear themselves to both students and coworkers. Because the death of a colleague can be devastating to teachers and staff, it is imperative to support them so they can better care for their students (Roth, 2015).

When a beloved kindergarten teacher died after a long battle with cancer, a first grade teacher invited a responding school psychologist into her classroom to help discuss her colleague's death. There was conversation about feelings, sadness, missing her, and then pleasant memories and lessons learned in her kindergarten class. The discourse expanded as the children shared stories about deaths of relatives and pets, affording the opportunity for *secondary triage* to identify those who might benefit from follow-up. The visit included bibliotherapy—a reading of Judith Viorst's *The Tenth Good Thing About Barney* (1972), a story about coping with the death of a pet cat. The children were reminded there were helpers in school with whom they could talk. As the responder left the classroom, the teacher managed a smile through her sorrow (Roth, 2015).

Context of the Death

The context of the death must be considered in assessing the degree of impact on the school community, which determines the *level of response* needed. Considerations influencing impact include:

- status of the deceased staff member in the community, including length of time working, and how well known and well liked;

- greater impact when family members of the deceased attend or work at the school;
- cause and manner of death—sudden deaths tend to have more intense reactions than expected deaths (terminal illness) that allow preparation and anticipatory grief;
- location of the death—occurring on campus, traveling to or from school, or during a school-related activity can be more traumatic than death off campus, especially for witnesses;
- schools that experienced previous trauma and grieving process may have prior emotions re-surface.

Source: Adapted from Poland et al., 2014

Grief

Reactions that may be experienced and re-experienced during grief:

- accepting the reality of death;
- experiencing the emotional pain of the death and missing the deceased;
- adjusting to changes—an environment that no longer includes the deceased;
- remembering and memorializing the deceased (Worden, 2009).

Source: Adapted from Fernandez et al., 2015

Reactions indicating possible need for additional assistance:

- significant loss of interest in daily activities;
- changes in eating and/or sleeping habits;
- wishing to be with the deceased person;
- fear of being alone or isolation and withdrawal;
- notable decrease in academic achievement or school attendance;
- increased somatic complaints;
- increased irritability, disruptive behavior, or aggression;
- ongoing, intense, intrusive, pervasive thoughts and feelings;
- severe maladaptive behaviors.

Source: Adapted from Fernandez et al., 2015

Avoid saying or doing the following when responding to grief:

- Avoid euphemisms for death such as "sleeping", "gone away" or "at rest"—use terms like "death" and "died" to avoid fear and confusion.
- Avoid minimizing statements such as "at least it was only your aunt" or "pet".
- Avoid imposing a time frame to complete the grieving process such as "you've grieved enough, you should feel better" or "it's time to move on".

- Avoid over-identifying statements such as "I know how you feel".
- Avoid too much self-disclosure—focus counseling on the grieving person.

Source: Adapted from Fernandez et al., 2015

PREVENTION AND PREPAREDNESS

The school principal and counselor embraced the task of mitigating the impact of the kindergarten teacher's imminent death by preparing the staff and comforting the already grieving family. The school counselor provided ongoing information, educational materials, and group discussion for teachers, who needed sufficient emotional stability to care for themselves and their students. Teachers and families in the school community were educated about the grieving process, children's typical reactions to death, how to reaffirm a sense of security, when to refer for more support, and how to use creative activities to facilitate self-expression. The principal kept the staff apprised of their colleague's struggle and ways to support her family (Roth, 2015).

While the teacher's death was anticipated, it was still a shock when she died. Working through their sadness, the teachers supported their students and each other. They used age-appropriate language to sensitively share the sad news. They planned ways to help their students remember the teacher and the lessons she had taught them. An unanticipated, sudden death would have precluded such extraordinary preparation, and reactions would likely have been even more intense (Roth, 2015).

School Bereavement Policy

Having a clear bereavement policy can provide a safe framework for school staff to respond to a death. A policy that includes proactive training and consultation enables staff to competently, confidently support bereaved colleagues and students. It can foster open communication about serious staff and student illnesses, thus enhancing preparation and support. A bereavement policy can include 1) procedures to carefully inform staff and students about a death, 2) a school crisis team with training to support staff and students under a variety of traumatic scenarios, 3) designated staff members to play support roles, 4) training and consultation for staff and families on how to recognize typical and more complicated grief reactions and behaviors, 5) procedures and roles addressing key tasks such as working with the media, compiling informational resources and books for bibliotherapy, and planning memorials, and 6) connections with district, community, and regional support in the event of a severe disaster with loss of life (Poland et al., 2014; Roth, 2017). The PREPaRE model is highly recommended as a comprehensive preparation and intervention framework (Brock et al., 2016).

School administrators working with school-based mental health professionals play a critical role in establishing policies, procedures, and norms of problem solving, professional development, recognizing and addressing needs, open communication and support among staff, and modeling and teaching the principles of resilience.

RESPONSE

See essential elements of crisis response outlined in the introduction to Section 2, Death of a Student, and in the **Introduction** to this book, *Essential Elements . . .*, which are relevant to the Death of a Teacher/Staff Member. The reader is also referred to "Suggestions for teachers and staff to support bereaved students" outlined in the introduction to Section 2.

The way schools respond to a death can help minimize emotional trauma on those affected and facilitate recovery and a return to a normal routine. Advance planning is required for an effective response. Schools offer an affected school community a familiar environment where supportive services can be provided for large numbers of people. Emotionally stabilized teachers can be a source of support, and trained crisis responders can provide triage and interventions.

Suggestions to support teachers and staff after a death:

- Stabilize teachers and staff, offering crisis counseling as needed.
- Provide script for classroom announcement and in-class support as needed, especially while discussing the death.
- Provide floating substitutes to relieve teachers needing counseling or quiet time.
- Provide substitutes or adjusted schedule to allow time for school staff wishing to attend funeral services.
- Provide individual counseling through the Employee Assistance Program as needed, and a support group if staff expresses a need.
- Consider a developmentally and culturally appropriate, life-affirming memorial, perhaps after hours or evening, that easily allows choice of voluntary attendance for students, families, and staff.

Parents/Guardians Can Support Adaptive Coping

Suggestions for parents to help children cope with death:

- Have a talking time to hear your child's thoughts and feelings. Understand your child's perspective to correct false beliefs. Don't force—follow your child's cues.
- Explain directly and age-appropriately what happened and answer questions honestly. Avoid unnecessary or disturbing details. Assure reasonably that you will care for your child.

- Be prepared for your child to raise difficult subjects. Look for cues indicating the need to be reassured, to better understand, or talk about details or feelings repeatedly. It is okay to say there are things we don't know.
- Be aware that you and your child may be in different places dealing with the death. You may reach a stage of anger or acceptance, while your child remains in a state of grief or sorrow.
- Correct clearly and directly any false assumptions about death, such as guilt or magical thinking that his/her thoughts or actions caused it.
- Be prepared to deal with the topic of death whenever your child brings it up, reassuring over a long period of time.
- Do not let your child watch disturbing images on television or internet. Minimize trauma exposure—spend time with your child away from television and media.
- Reestablish your child's daily routines—especially bedtime.
- Family religious or spiritual beliefs can be integrated into thoughts, discussions, and ways to find meaning and cope with the death.
- Let your child write a letter, draw, or in some way express sympathy for those grieving. Older children may find comfort in writing their thoughts and feelings. Teenagers often benefit from guided peer support or life-affirming activities.
- Expect increased need for physical contact and fear that something will happen to you or other loved ones. Expect more sadness, clinging, anger, or playfulness.
- Grief reactions are normal, but if your child continues to experience severe distress or has difficulty with normal functioning, seek help from professional resources in the school or the community.

Source: Adapted from Raundalen & Dyregrov (2004)

Suggestions for parents to help teenagers cope with death:

- Be a listener. Reflect genuine concern without giving advice.
- Share that grieving may last longer than expected, though the intensity usually subsides over time. Encourage healthy expression of grief.
- Do not avoid talking about a person or people who died or the event, fearing it might re-awaken the pain. Usually, teenagers want to talk, but in a manner and a time they select. Follow their lead.
- Times of grief are not times to make changes or important decisions. Try to keep the situation as normal as possible.
- Watch for trouble signs in adolescents. The need to appear competent or "cool" may prevent teenagers from reaching out for assistance.
- While most grief reactions are normal and temporary responses to death, it is imperative that suicidal thinking be treated seriously, and that help is sought (Erbacher et al., 2015).

- Referral should be considered if trouble signs are especially severe and intense over an extended period, or represent striking changes in usual behavior. Trouble signs may include:

  - withdrawal and isolation;
  - physical complaints (headache, stomach pain, insomnia);
  - emotional concerns (depression, anxiety, suicidal thoughts, confusion);
  - anti-social behavior (stealing, aggression, alcohol or substance abuse);
  - school problems (avoidance, disruptive behavior, academic failure).

- The power of the peer group is often evident when dealing with teenage grief. Adolescents may form networks of support, leaning on each other during difficult times. Encourage helpful, cooperative, life-affirming activities.
- Less frequently, the peer group can be a catalyst for further tragedy. After a suicide, caregivers must be vigilant to prevent suicide contagion, clusters, or pacts that signal the danger of further suicides among those most at risk.
- While most teens support each other moving toward recovery, some may succumb to suicidal thoughts or intent. This emphasizes the need for triage, discussion, psychoeducation, constructive actions to empower, and when necessary, monitoring and referral for more intensive treatment.

Source: Adapted from *Responding to Critical Incidents:
A Resource Guide for Schools*. British Columbia
Ministry of Education (1998)

RECOVERY

See points of recovery emphasized in Section 2, Death of a Student, and reference to the "Introduction" of this book, *Essential Elements . . .* Disengagement: Final Phase of Response, which are relevant to the Death of a Teacher/Staff Member.

REFERENCES

British Columbia Ministry of Education. (1998). *Responding to critical incidents: A resource guide for schools*. British Columbia, Canada: Author

Brock, S.E., Nickerson, A.B., Louvar Reeves, M.A., Conolly, C.A., Jimerson, S.R., Persce, R.C., & Lazzaro, B.R. (2016). *School crisis prevention and intervention: The PREPaRE model* (2nd ed.). Bethesda, MD: National Association of School Psychologists.

Erbacher, T.A., Singer, J.B., & Poland, S. (2015). *Suicide in schools: A practitioner's guide to multi-level prevention, assessment, intervention, and postvention*. New York: Routledge.

Fernandez, B., Comerchero, V.A., Brown, J.A., & Woahn, C. (2015). *Addressing grief: Tips for teachers and administrators*. Bethesda, MD: National Association of School Psychologists.

Poland, S., Samuel-Barrett, C., & Waguespack, A., (2014). Best practices for responding to death in the school community. In P. L. Harrison & A. Thomas (Eds.) *Best practices in school psychology: Systems-level services*. (pp. 302–320). Bethesda, MD: National Association of School Psychologists.

Raundalen, M. & Dyregrov, A. (2004). *Terror: How to talk to children*. Center for Crisis Psychology, Bergen, Norway. www.icisf.org/articles

Roth, J.C. (2015). *School Crisis Response: Reflections of a Team Leader*. Wilmington, DE: Hickory Run Press.

Roth, J.C. (2017). The importance of consultation in supporting bereaved students. In J.A. Brown & S.E. Jimerson (Eds.), *Supporting bereaved students at school* (pp. 52–69). Oxford, UK: Oxford University Press.

Viorst, J. (1972). *The tenth good thing about Barney*. New York: Atheneum.

Worden, J.W. (2009). *Grief counseling and grief therapy: A handbook for the mental health practitioner* (4th ed.). New York: Springer.

5 Death of an ELL Teacher

School Researcher Experiences a Crisis

Sara M. Castro-Olivo

INTRODUCTION

The crisis I will describe in this narrative took place during my first year as an assistant professor at a research institution in Southern California. I was fortunate to have found a few schools that allowed me to implement a culturally adapted Social-Emotional Learning (SEL) program to their Spanish-speaking English Language Learners (ELLs). One of the schools I worked with was in a small town less than 100 miles East of Los Angeles. The school district was predominately Hispanic (60% of students identified as Latino and 25% ELLs [California Department of Education, 2016]). I was very excited about this partnership. At the time, I was conducting a study that evaluated the effectiveness of a culturally adapted 12 lesson SEL program for Latino ELLs enrolled in 6th–12th grade. In the fall of that year, I met with many educators to explain my study/intervention. I often shared my own experiences as a former ELL student and why I felt the need to teach English Language Learners SEL and coping skills. I was welcomed right away by many teachers and ELL directors. At one high school, the ELL coordinator used to co-teach the ELL class for new students (ELL I) with a vibrant and friendly teacher. The teacher, named Anna for this narrative, shared that she had been an ELL as she was an immigrant from Russia, and understood the social-emotional needs of her students. The ELL coordinator was a bilingual Mexican-American young professional, named Rocio for the purposes of this narrative. Rocio was very passionate and determined to give her students the best educational experiences possible. They were both enthusiastic to have their students be part of my study and seemed more than happy to help in any way they could.

I was very excited to work with these two amazing ladies and their newcomer ELL students. The classroom was mixed (9th–12th graders) and had two languages represented. Most the students had been in the U.S. for less than one year. Because of the students' limited English language skills, the program was delivered in Spanish for the native Spanish speakers (22 of them) and translated into Tagalog by a teacher assistant for the two Filipino students in the class. Given the need to implement the program in Spanish, Rocio volunteered to teach it. I provided the training, supervision and on-going consultation. We began the study in February. I visited the classroom on a weekly basis to conduct fidelity checks and provide consultation and feedback to the teacher (Rocio). The students got to know me. I explained the study to them, and shared my story as an ELL. They saw me on a weekly basis in the back of their classroom providing support to their beloved teacher Rocio. At the beginning of the 12-week intervention, Anna sat in the front of the classroom and attentively listened to the SEL lessons. She provided examples whenever possible. It was clear she understood some Spanish and was eager to help her students in any way she could. At that time, students seemed to respect her, but it was clear to me that they felt closer to Rocio. At first, I thought the relationship differences between the students and the two teachers was due to language and cultural congruity. Soon I found out that there was another explanation.

Anna, the lead teacher for the class, seemed to be disengaging from her students and colleagues. It seemed Rocio was getting frustrated that she had to take over the classroom more and more. Still, she was willing to do anything to make sure her students were not left behind. Rocio was only supposed to be there to support the class and Anna, but as the weeks went by, the class seemed more Rocio's than Anna's. I didn't say much, thinking Anna was probably choosing to disengage so the students could connect with someone who looked and spoke like them. To some extent it seemed natural to me at the time.

I became concerned when Anna's withdrawn behavior started to turn into absences and later, when she seemed to be coming to work intoxicated. She was unkempt, with no make-up, messy hair, and looked as if she had not slept well. She appeared tired and cognitively unable to maintain a conversation. Soon the students' nonverbal language demonstrated frustration and even lack of respect for Anna. Rocio experienced similar emotions. She was no longer the supportive, proud colleague who had introduced Anna as an amazing and dedicated teacher in the beginning of the year. She seemed upset. This all happened during the first 6 weeks of my study. I kept my observations to myself. I was an outsider and did not feel comfortable pointing out my concerns to Rocio or Anna. I was afraid to be viewed as unprofessional or insensitive. I feared that if I talked to Rocio, I would be viewed as wanting to start gossip. If I talked to Anna, she might think I was rude.

LACK OF PREPAREDNESS

A Day of Crisis

We were scheduled for our eighth session that morning. I received a text from Rocio at 5 in the morning informing me that her principal had just called with the news of Anna's passing. She had been battling cancer. Rocio was devastated. She could barely speak and every other word that she uttered that morning indicated that she felt guilty for not knowing what was going on and not being a "better" and more supportive colleague to Anna. It was clear to me she was inundated with guilt and sorrow.

No one at the school was aware of Anna's battle with cancer. She had requested a personal leave for the last two weeks and never came back. Students, teachers and administrators were surprised to find out that she was sick and had been suffering for some time from a terminal disease. Anna must have been in her early forties. She seemed full of life at the beginning of the semester. As mentioned, toward her last weeks she seemed like someone who had a night full of fun, still managing to smile while looking intoxicated and not well groomed. I also felt guilty when I found out what she had really been going through. If I had only known? That's what we all said to ourselves. It was so hard not to judge Anna for the way in which she would show up at school.

Rocio's request that morning was clear, "I am devastated and I am sure our students will be too once they find out . . . and we don't have any counselors who speak Spanish at our school to help them with those feelings. We need you, please come!" she said. I tried to process everything as soon as I could and got myself ready. I got to the school before classes started. During my 30-minute drive to the school, all I could think about was, why didn't I approach Anna? How did she feel? How does Rocio feel? She must be in worse shape than the students . . . so many thoughts came to my mind that I couldn't even think about recommendations for the crisis team.

DESCRIPTION OF RESPONSE

The Crisis Team and Response for Students

When I got to the school, a crisis team had already arrived with two other bilingual individuals the district had contracted. The school was set to go. They had already written the statement to be read to students. The letter that was going to be sent home to parents was being reviewed by the district's legal office. The crisis team determined that group grief counseling for those students who needed it, would take place at the library after the announcement and during the scheduled class period students were supposed to have with Ms. Anna. They made the announcement at the beginning of the day in the gym. The five classes Ms. Anna was teaching were asked to report to the gym first thing in the morning. The rest of the school was informed via intercom after her students were informed at the gym. The set up didn't seem appropriate to me, but as an outsider to the crisis team, I didn't feel comfortable speaking up. I wish I could have had time to talk to the crisis team leaders and the school's administrators to suggest that ideally, students should be informed of traumatic events in their classrooms, where their reactions can be better monitored, and referrals for those needing crisis counseling could be better managed. That's what experts like Brock and colleagues (2016) suggest, and is what I have been trained to do. Being able to monitor the students' reactions would have allowed for better triage and services for students who were more affected by the news. I could tell all students were affected, but not all of them were able to process their feelings the same way. While concerned about the procedures, as an outsider I felt at that point my place was just to listen and support in any way they needed me after the news was shared.

I made it a point to be next to Rocio the entire time. She was very affected, but was working hard to stay calm and try to support her students. It was clear to me that the rest of the crisis team was only focused on the students and neglected to check on teachers and staff members. Ms. Anna's teaching assistants/interpreters were not even in the room. I cannot recall asking about them. I have since learned that best practice emphasizes the importance of emotionally stabilizing and supporting staff so they are better able to care for their students (Brock et al., 2016). I wish I could have been more vocal about this issue at that point. It felt natural for me to check on Rocio, but what about the rest? After the news was shared, only the contracted counselors, Rocio and I stayed with the students who chose to stay in the library. I wondered how everyone else was coping with the news and how teachers with students who had just been told were dealing with them in their classrooms.

Once the principal started addressing the students by reading the written statement, I was shocked to see that the interpreter translating for the Spanish-speaking students was not provided a copy of the statement before hand. He translated as the principal spoke, and to my judgment, made a few poor choices of words when talking to the students. I am not sure if more sensitive language would have made any difference, but as someone who had also been affected by the news, feeling guilt and sorrow for having judged Anna and not figured out a way to support her, I wish he had been given the opportunity to see the statement and pre-select more sensitive words. Rhodes, Ochoa and Ortiz (2005) suggest that interpreters need to be highly trained in psychoeducational terms prior to being asked to interpret. It is also suggested that prior to interpreting for a live audience, interpreters should be given an explanation of the materials to translate, the sensitivity of the issue to be addressed, and time to ask questions if the interpreter expects any potential issue related to the cultural sensitivity and/or competence of the text to be translated. None of these practices seem to have been addressed during this incident. I felt bad for the interpreter. I could tell he was also impacted by the news and was struggling with sharing it and having to translate on the spot during a time of distress for him.

By the time the news had been shared, many kids were in tears and took the option to go to the library for crisis counseling. They had 6–8 tables at the library. Each table had some paper and markers for the students to write letters or draw pictures they wanted to share with Ms. Anna's family. Crisis intervention counselors walked around the tables where most of the students seemed to be processing feelings on their own, writing letters and using craft materials.

I started going to each table but spent more time at tables with kids who were part of my study. It was not intentional, but the students who knew me from the study seemed more open to me and willing to share their feelings and ways they were processing them. I also felt a stronger need to connect with these students. Maybe it was my own behavior that made them open up more —or a combination of them knowing me and me knowing them. In retrospect, the sense of responsibility I felt for the students who had participated in my study made me realize the importance of having crisis experts, or teams, as part of every school, because of feeling more connected when they are *your* students.

I was surprised to see the level of resiliency these kids were demonstrating. The guilt theme kept emerging from table to table, but they were truly utilizing the skills they had learned as part of the SEL program. They used specific vocabulary words we had introduced in the program and gave me a few examples on how they were "reframing" their grief and guilt. I was proud to see how the students who were participating in my study were applying the skills, but deeply saddened to see that not all the tables were engaging in the same level of discussion or even any discussion at all. This experience also helped me realize the importance of teaching SEL programs to all students so they can be more fluent in emotional language for situations like this. All students deserve the basic emotional literacy to engage in these constructive conversations when they are needed, and to know that it is okay to become emotional and process these feelings with others. Social Emotional Learning has been defined as skills that allow us to understand and manage our emotions, set and achieve positive goals, show empathy for others, establish and maintain positive relationships, and make responsible decisions (CASEL, 2016). In the program I was evaluating prior to the incident, we had taught students self awareness, social awareness, empathy, problem solving, anger management, cognitive restructuring and positive thinking. Being aware of their own feelings truly helped the students identify thinking errors and reframe those errors for more positive thoughts and behaviors that allowed them to better cope with the situation.

STRIVING FOR RECOVERY AND MEANING

Personal Feelings

As previously mentioned, I was struggling with the death of this teacher too, as I had worked with her and internally criticized her behavior during her last weeks of life. I had no idea she was dying and being at school though she looked intoxicated was probably her only way to cope with an inescapable reality. There was so much I wish I could have told her. I wish I would have gently confronted her the many times her behavior was unprofessional and leading to disengagement from her students. A simple "are you okay, are you sure you are okay, you seem tired and not ready to teach?" Although intrusive, this could have led to sharing her situation and we could have all prepared better for her last days. It was traumatic for everyone, and she left without us being able to tell her something nice. She was truly liked by her students and colleagues. It was so sad that she left during a time people had been doubting her due to behaviors she only engaged in during the final few weeks.

Many mentioned that they wished they had told her how much we admire her for battling her illness with dignity and trying to protect those around her. We guessed that she didn't inform others about her illness because she did not want her students to suffer with her. Perhaps she felt that suffering alone was enough, or maybe she did not feel strong enough to deal with her feelings and her students' feelings. Who knows the real reason why she kept the news from everyone? Respecting her privacy and right to not disclose is something we all have to think about, but it was surely hard to deal with such a shock. I guess a sudden death from an accident or event without warning signs that something was wrong would have been different. We all seemed to be struggling with the fact that we did not know about her battle and had criticized her for coming to work "that way" more than the fact that she was now resting in peace. She only missed the last two weeks of school. Everything happened so quickly! We were so overwhelmed. Sad and guilty.

LESSONS LEARNED AND REFLECTIONS

Preparing Communities for Loss

Many thoughts came to mind after the situation. First was the need to work with teachers about ways to prepare themselves and their communities about potential deaths. I believe teachers should have the right to share as much (or little) as they feel comfortable, but it is important to help them make an informed decision about preparing their students for an inevitable loss. Students seemed to have been more affected by the fact that they did not know what their teacher had gone through than her actual death. It seemed to have added an additional layer to their grief. I am sure the adults (teachers, staff and administrators) felt the same way. I think it is important to have an open conversation with all teachers about the effects their death or the death of colleagues could have on their students. School staff could be encouraged to work with school-based mental health providers to identify ways to share information about their own serious illness, or support students if a colleague is seriously ill or dies. Teachers should feel comfortable enough to share information about their illness with administrators and mental health care providers, and together prepare themselves and their communities for a potential loss.

Preparing Crisis Teams on Culturally Responsive Practices

Equally important to preparing communities for potential loss, is the need to prepare crisis teams for culturally responsive practices. Providing culturally responsive crisis interventions facilitates individuals' grief in a way that feels more natural to each member of our diverse communities. Ortiz and Voutsinas (2012) make some insightful recommendations for culturally competent practices and remind us that, as school psychologists, we are ethically bound to strive for cultural competence in all we do, including crisis intervention. Being aware of the different ways members of culturally and linguistically diverse (CLD) groups process grief and death is very important in planning and conducting crisis response. In the present example, some students needed different materials (i.e. besides paper and color pencils for letters) and practitioners (i.e. bilingual and bicultural) to help them deal more effectively with their loss. It was also clear that a trained, school-based crisis intervention team could have prepared for a more culturally competent response and connected more easily with their bereaved students.

The Role of SEL in Crisis Intervention

As previously mentioned, although I was not able to quantify, I noticed a difference in the groups who had been exposed to SEL skills (i.e. self awareness, social awareness, empathy, self-management, cognitive restructuring, and positive thinking) than those who had not. They were more comfortable sharing their feelings. The emotions they experienced and shared seemed to be appropriate for the situation. In addition, they seemed to be processing their guilt and grief at a faster rate than students seated at tables where no one had been part of our SEL study. Students who had been in our SEL study appeared to move toward expressing appreciation and making tributes to their teacher at a faster rate than those students who had not been part of our study. They spent less time "quietly" dealing with their own feelings. More research is needed in this area, but this experience highly motivated me to continue promoting SEL programming in all classrooms to ensure all students have the basic emotional literacy skills and ability to identify and manage their feelings, even during unexpected traumatic moments.

Concluding Lessons and Personal Reflections

This experience marked my career. It helped me realize the importance of universal SEL interventions. It also made me think about the need to have crisis interventionists be part of the community they serve. It seemed these students were more open to talking with someone they knew versus someone they didn't know. As mental health care providers, we need to make a point to be visible in our schools. Co-leading SEL efforts at the universal level seems to be the most appropriate and logical way to ensure we are providing students with the skills and support they will need in any situation. This experience also emphasized the need to always practice from a culturally responsive lens. Our schools are very diverse settings and not all students cope with grief the same way. It is extremely important for practitioners to be very familiar with the communities/groups they serve and be as culturally aware as possible about the role culture and language play in the way students and staff process their grief. Using "one" wrong word could lead to a group of students feeling alienated, and this could lead to an even stronger, less adaptive grief reaction. Working with cultural liaisons and continuously working towards culturally competence should be our number one goal.

REFERENCES

Brock, S.E., Nickerson, A.B., Reeves, M.A.L., Conolly, C.N., Jimerson, S.R., Pesce, R.C., & Lazzaro, B.R. (2016). *School crisis prevention and intervention: The PREPaRE model* (2nd ed.). Bethesda, MD: National Association of School Psychologists.

California Department of Education (2016). QuickQuest: Demographic Data File. Retrieved from: http://dq.cde.ca.gov/dataquest/content.asp.

Collaborative for Academic and Social Emotional Learning (CASEL, 2016). What is SEL? Retrieved from: www.casel.org/what-is-sel/

Ortiz, S.O., & Voutsinas, M. (2012). Cultural Considerations in Crisis Intervention. In S.E. Brock & Jimerson, S.R. *Best Practices in School Crisis Prevention and Intervention* (2nd ed.). Bethesda, MD: NASP Publications.

Rhodes, R., Ochoa, S.H., Ortiz, S. (2005). *Assessing culturally and linguistically diverse students: A practical guide*. New York: Guilford Press.

6 Cancer Sucks
Deaths of a Teacher and a Nurse

Christina Conolly

INTRODUCTION

One fall Saturday morning, I received a phone call that a teacher from one of the elementary schools had died of cancer. The teacher was beloved by staff at the school and throughout the district. She had been on extended leave from the start of the school year and was frequently in and out of remission from cancer. The school had fundraisers on her behalf and large assemblies when she returned after being ill for long periods of time. The teacher lived in the school community, sat on her front porch when students walked home from school, and spoke with them daily. The building administration was very close with the teacher and was extremely upset when she died.

As I prepared for crisis response on Monday, I got news that a staff member on extended leave from another building to receive cancer treatment died. She was a middle school nurse, well known to the building staff and all of the nurses in the district. Students who received services from her during previous school years were also close with her. The nurse had been away from the school the majority of the previous school year, so many students did not know her. This was the second staff member death at the school in approximately three years.

DESCRIPTION OF RESPONSES

On Monday, students were not attending due to a teacher in-service day. An Incident Action Plan (IAP) was developed to assist the district crisis team respond to both incidents (Brock et al., 2016; Reeves et al., 2011). The IAP was written to provide guidance to members of the district crisis team regarding their specific duties during the response. Building administrators assisted in the development of the response plan. The district crisis team included approximately 30 mental health professionals (school counselors, psychologists, and social workers) split into three teams. Team A was assigned to provide assistance to the middle school. Team B was assigned to provide assistance to the elementary school. Team C was on call to provide assistance to the elementary school as needed.

Middle School Response

Negotiating with School Administration

The middle school building administration shared the opinion that a large crisis response was not needed. The administration initially did not want *any* crisis response at their building. They were concerned that there would be an increase in staff and student distress if the crisis team responded in the building. There was a perception that staff should not display crisis stress reactions and school should continue as normal. After some convincing, the administration allowed Team A to come to the building to provide support for the staff. *Caregiver Training* was provided to the staff during the day of in-service (Brock et al., 2016).

It was difficult to triage the individuals who were emotionally proximate or had previous vulnerability factors that might indicate the need for intervention (Brock et al., 2016). The administration was not forthcoming with information needed to determine who might be most impacted by the death of the nurse. I received some information from the head of district Health Services, but the building staff did not know of many students who were emotionally close to the deceased nurse. Many students the nurse worked with had graduated from the building the previous spring. The nurse was also gone for most of the previous school year due to her medical condition. Students and staff knew that the nurse was terminally ill and not likely to return.

Support for School Staff

On Tuesday, members from Team A were stationed in a conference room to provide support for staff as needed. A light breakfast and lunch were provided in the conference room for staff who came for support. Although the culture of the building discouraged expressing stress reactions, staff members slowly came throughout the day to eat the food and receive individual crisis intervention with a response team member. Many staff members shared that the food provided them the "excuse" to receive support from the response team. Two members from Team A were available to conduct *Psychoeducational Groups* in classrooms after the announcement of the death was made in each class (Brock et al., 2016). The majority of support was needed for the building staff.

District-Level Nurse and Staff Response

During the staff in-service on Monday, members of the district crisis team met with the nurses during their training. Two members of the team provided *Caregiver Training* to support the nurses. Another crisis team member floated between the health insurance and human resources departments to provide support for district-level staff who had interacted with the deceased staff members during their cancer treatments over the years.

The following day, all district staff members were invited to attend a support session in the afternoon after school, if they felt the need. Many district staff members knew both the teacher and the nurse and worked with them for years. This was an informal meeting where staff could come together, reestablish social support, and grieve together. Two mental health providers on the district response team were assigned to walk around during the session and provide support as needed. Food was also available to help ensure that basic needs were met (Brock et al., 2016).

Elementary School Response

Significant Emotional Distress

The elementary school involved a multi-day crisis response. The majority of building staff, including the administration, demonstrated significant stress reactions. The district crisis team utilized both Teams B and C to provide support. The majority of students and staff experienced high levels of emotional distress. The school-based mental health providers could not offer sufficient support to others due to their own levels of emotional proximity to the teacher. The building administrators were so distraught that district administration was concerned they might be temporarily unable to run their school without support. One of the school administrators placed a large sign outside of the building during the weekend, informing the small school community of the teacher's death. As students and staff walked into the school Monday and Tuesday mornings, it was the first thing they saw, increasing the amount of staff and students with significant emotional reactions.

Administrators Feel the Burden

When we respond to crisis events, we must always remember that we work with human beings who feel strong emotions about traumatic incidents in their schools. Often, we forget that our building administrators may experience considerable grief when a student or a staff member dies. Even the most capable administrators can be impacted significantly when someone close to them dies. The administration at this building, a highly effective administrative team, were no exception to experiencing the pain and suffering that comes when someone you care about dies. Administrators often assume the greatest responsibility and heaviest burden for their building, and need district-level response team support.

Problematic Decisions Escalate Crisis Reactions

The Monday of the staff in-service, members of Team B had worked with the staff members and provided *Caregiver Training* (Reeves et al., 2012; Brock et al., 2016). Toward the end of training, one of the building administrators decided to show the staff a video of the deceased teacher filmed at the school. The video stirred emotions to the extent that the Caregiver Training intervention soon turned into a more intensive *Group Crisis Intervention* (Brock et al., 2016). Members from Team B had to "re-group" and allow staff members to share their stories about the teacher and their reactions toward the death. The crisis team then discussed stress management techniques with the staff and what the crisis response would look like when students returned the next day. In sharp contrast, the crisis interventions provided at the middle school took approximately 30 minutes, while the supports provided at the elementary school lasted the majority of the in-service day.

Interventions to Support and Stabilize Staff and Students

Several other administrative decisions that may not have been in the best interest of the response were influenced by the building administrators' grief reactions. Multiple district administrators were concerned about these decisions and asked me to monitor the crisis response. The school administration was informed that starting Tuesday, any building level decisions regarding the crisis situation were first to be discussed with, and approved by me before being implemented. Although I previously decided to temporarily relieve teachers and school mental health professionals from caregiver roles, I was not tasked to determine if an *administrator* should be temporarily removed from the caregiver role. I was nervous about this responsibility, but luckily I had good rapport with the building administration prior to the incident. I provided activities for the building administrators such as contacting families of students who were impacted by the teacher's death. Since this was a small, neighborhood elementary school (300–400 students) the community was very close and families appreciated hearing directly from one of the building administrators. A building administrator also worked with the building and district Public Information Officers (PIO) to develop a letter to families that discussed the crisis intervention activities at school and the guidelines for students and families choosing to attend the funeral. Providing concrete, proactive tasks helped the administrators stay focused on the communication needs and supports that the families required.

After reviewing the response needs during the in-service, I worked with Human Resources to request a few substitute teachers assigned to the school to fill in for teachers who were unable to teach either due to significant crisis reactions and/or wanting to receive *Individual Crisis Intervention* (Brock et al., 2016). Team C, who had been on stand-by if Team B needed additional mental health staff provided that support. At this moment, all members of the district crisis team were called in to respond to both buildings.

Members from Teams B and C were assigned to conduct *Psychoeducational Groups, Individual,* and *Group Crisis Intervention* (Brock et al., 2016).

Psychoeducational Groups were conducted in those classrooms experiencing moderate to high levels of emotional proximity to the teacher who died—primarily the third to fifth grade classrooms. Less intense *Classroom Meetings* were conducted with the kindergarten, first, and second grade classrooms due to their lower levels of emotional proximity (Brock et al., 2016). Students who needed additional support were referred to either group or individual interventions as needed.

As members of the crisis team provided support to the upper grade levels, extra support was provided to the third grade classroom assigned to the teacher who died. Although a long-term substitute teacher was with the class since the first day of school, their ailing teacher was in contact with them during the first few weeks of school. The students knew she was their teacher. She had also been in constant contact with the long-term substitute. Two members of the mental health response team went to initially provide a Psychoeducational Group, but quickly transitioned into a more intensive Group Crisis Intervention once high levels of emotional trauma became apparent (Brock et al., 2016). The long-term substitute also displayed high levels of emotional proximity. Crisis team members initially asked if she wanted to receive private, individual support when they worked with the students. She declined, wanting to be with her students. As the team members talked with the students and they shared their stories, the teacher became upset and left the room. She came to the incident command center and requested to talk with someone. She informed me that she initially thought that she could handle it. The day before, she and other staff had a whole day to process everything. She said that once she heard the students talk, it brought all of her emotions to the surface again. She then received an individual intervention and was able to return to her class before the team members completed the group intervention.

RECOVERY

Members of the district crisis team continued to support the building for a couple more days to follow-up with students and staff needing more attention. It was critical for the district team to continue supporting students and staff since the building caregivers were not yet able to conduct follow-ups due to their own crisis reactions (Brock et al., 2016). I continued to monitor activities and support the school administration. Once I left, the principal's supervisor was there to help as needed. I continued to provide support and consultation for the next two weeks.

Support During the Funerals

Another consideration was providing support during the funerals. The funeral for the nurse was on the weekend. The funeral for the teacher was during the school day. We received approval to provide substitutes for the elementary school staff that wanted to attend the funeral. We also received approval to assign district-level administrators (with appropriate teacher certifications) to help cover classrooms while the staff went to the funeral. Parents were told that they could take their children to the funeral with a written note informing the school that they were attending. I was at the school providing coverage for the building administration and monitoring the crisis reactions of the students and staff who did not attend the funeral. Members of the crisis team were also at the school monitoring for reactions.

SPECIAL CONSIDERATIONS

Responding to Multiple Incidents

In responding to these crises, there are two special considerations that "pop out" of my reflections. The first consideration is the ability of districts to respond to multiple crisis events taking place at different buildings on the same days. It can be tough for a crisis team to respond to one building after a crisis event. Responding to multiple events at multiple buildings can become very stressful if a team does not have a well-coordinated system in place. Schools across the nation are encouraged by the U.S. Department of Education (US DOE) and the Federal Emergency Management Agency (FEMA) to incorporate the National Incident Management System (NIMS) and the Incident Command System (ICS) into crisis response capability (Brock et al., 2016; Conolly-Wilson & Reeves, 2013; Reeves et al., 2012; U.S. Department of Homeland Security, 2008). The ICS helped our district crisis team respond to two crisis events that impacted three buildings. I was designated district-level Incident Commander and there were two building-level Incident Commanders (one at each school). The central office was managed by the district-level Incident Commander. Key members of the ICS team included the Incident Commanders, Public Information Officer (PIO), Planning Section Chief, Operations Section Chief, and the Logistics Section Chief (U.S. Department of Homeland Security, 2008).

At the middle school, the building Incident Commander did not delegate responsibilities to any members of the building's ICS team. At the elementary school, the building PIO and the building Incident Commander assisted the district PIO in preparing information to send out to the community. Other members of the building ICS team were too impacted by the crisis event to respond. Under the district Operations Section, the district Mental Health Response Team was able to split into three groups to provide support to multiple buildings. The ability of the district ICS team to expand and contract as services were needed was critical to providing the support that the buildings needed. In addition, the district-level Logistics Section was able to provide all of the necessary equipment that was needed to ensure good communication between the three buildings. The district utilized a digital radio system that included a district-wide radio channel. This channel allowed staff to communicate with each other across multiple buildings. This allowed me to stay in the building that had the most need, but still learn about responses at the other two buildings in real time.

Supporting and Stabilizing Affected Staff: Issues of Authority

A second consideration is how to provide support to a building when the primary caregivers, including administrators or mental health staff, are impacted by the crisis event to the extent that they cannot function effectively. When we examine how to *Care for the Caregiver*, a critical procedure to consider is when to remove a caregiver who cannot provide support (Brock et al., 2016). Luckily, I had the confidence of the Superintendent and members of Cabinet to consult with, and if necessary, make the decision to temporarily remove affected caregivers. That was not an easy task. Working in a district where I had rapport with the staff helped me tremendously. I cannot imagine what it would have been like if I had not known the staff. The Human Resources Department also allocated substitute teachers to allow teaching staff to leave their classrooms and receive support, which was critical to helping the staff grieve and recover.

It is suggested that procedures be developed for involuntarily removing a staff member from a caregiver role before, or during the crisis event. The plan should include who has the authority to decide to relieve a staff member if deemed necessary, and when the person can return to the caregiver role. Human Resources and possibly members of the staff and union associations should be involved in plan development. However, it is important to keep in mind that a critical role of the district-level crisis team is to stabilize and support affected school staff so that they can maintain the caregiver role, or return to it as soon as possible (Brock et al., 2016). Often, consultation with the principal and school mental health providers, and blending the school and district response teams fosters resilience and recovery.

LESSONS LEARNED AND REFLECTIONS

When a school responds to the death of a staff person, there is potential to impact the entire school community. Educators affect the lives of students and each other on multiple levels. They can work in a school for their entire career and work with not only students, but with parents and families. When multiple staff members die in close proximity, a school district must have the mental health resources to respond in multiple buildings. Crisis teams must remember how to provide care for everyone in the school population, including administrators, teachers, and the mental health responders that work in the school. A district-level Crisis Team must have the resources to respond to multiple crisis events that occur simultaneously. Using the ICS team structure is an excellent tool to appropriately manage a response that is occurring at a single or multiple buildings. Although providing a response between multiple sites can stretch the resources across a district, having school-level and district-level teams properly trained in mental health crisis response and NIMS can greatly assist this mission.

REFERENCES

Brock, S.E., Nickerson, A.B., Reeves, M.A.L., Conolly, C.N., Jimerson, S.R., Pesce, R.C., & Lazzaro, B.R. (2016). *School crisis prevention and intervention: The PREPaRE model* (2nd ed.). Bethesda, MD: National Association of School Psychologists.

Conolly-Wilson, C.N., & Reeves, M.A. (2013). School safety and crisis planning considerations for school psychologists. *Communiqué, 41*(6), 16–17. Retrieved from: www.nasponline.org/publications/cq/index.aspx?vol=41&issue=6

Reeves, M., Conolly-Wilson, C., Pesce, R., Lazzaro, B., Nickerson, A., Jimerson, S., Feinberg, T., Lieberman, R. & Brock, S.E. (2012). Providing the comprehensive school crisis response. In S.E. Brock & S.R. Jimerson (Eds.), *Best practices in school crisis prevention & intervention* (2nd ed.). Bethesda, MD: National Association of School Psychologists.

Reeves, M., Nickerson, Conolly-Wilson, C., Susan, M., Lazzaro, B., Jimerson, S., & Pesce, R. (2011). *Crisis prevention and preparedness: Comprehensive school safety planning* (2nd ed.). Bethesda, MD. National Association of School Psychologists.

U.S. Department of Homeland Security (DHS). (2008, December). *National incident management system.* Washington, DC: Author. Retrieved from: www.fema.gov/pdf/emergency/nims/NIMS_core.pdf

7 Illness and Death of an Elementary School Principal

Melissa Heath

INTRODUCTION

When I think about responding to a crisis, one incident stands out as a particularly challenging experience. I was working in a Texas suburban elementary school. I had been a school psychologist for about four years, so I was not brand new, but definitely still learning about crisis intervention. It was in the mid-1990's, so the field of school-based crisis intervention was not well established.

My preparation to assist with crisis intervention did not occur in a classroom, nor did I receive training during school in-service meetings. I learned on-the-job. With minimal preparation, I sought help from my supervisors and found ways to address crisis situations as they arose. I was also fortunate that the district I worked with had several experienced school psychologists who had extensive experience with school-based crisis intervention. They formed a district team that helped when emergency situations arose.

Facing a Principal's Life-Threatening Illness

Ms. Hill, the principal of Shoshone Elementary School was previously a teacher at the school for a number of years prior to being appointed as the principal. In total, she had served at this elementary school for over 30 years. To say that she was loved does not fully capture the incredible respect and trust the school community felt for her. She was a grandmotherly, petite, and soft-spoken woman who was always at the school. No matter how early I came before school, she was there before me. When I left, no matter how late, she was always there meeting with parents, talking on the phone, planning academic interventions, or brainstorming school-wide activities with the vice principal.

Never married, her passion and all of her energy were funneled into Shoshone Elementary School. Students, teachers, and parents were loved and nurtured by this amazing principal. Because she never missed a day, everyone was puzzled when she did not come to work on certain days and when she began taking a few hours for "appointments." She seemed tired and distracted. None of us could believe that she was starting to slow down or that she might be ill.

With great sadness, she told Mrs. Clark, the vice principal, that the recent absences and appointments were because of an illness. However this illness was not merely the flu, but was recently confirmed to be leukemia. In confidence, Mrs. Clark told me to keep this news confidential because Ms. Hill wanted to announce her illness in her own way.

It was almost as if someone told me that the sun would not be rising tomorrow. I was in absolute disbelief. Over the next few days, Mrs. Clark and I had regular conversations, as we processed how we were dealing with the news and how we could help our teachers, students, and families through this challenging and uncertain time. Our biggest question and the question we anticipated from the children: "Was she going to die?" We did not have a definitive answer. For me, I can deal with certainties, no matter how harsh, but having our principal teetering on the edge of life and death—this was unsettling.

PREPAREDNESS

Informing Others about the Principal's Illness

When we first found out about Ms. Hill's illness, teachers and staff were preparing for Thanksgiving break. Children's turkey drawings with their traced hands for tail feathers were hung up and down the halls, classrooms were enjoying a Pilgrims' feast, and everyone could feel the excitement that comes right before a holiday. After school on the Monday before Thanksgiving break, Ms. Hill announced a special staff meeting for all teachers, staff, and paraprofessionals. She invited Mrs. Clark, several counselors, and me to be her backup support during and following the meeting. As anticipated, many responded with tears. This was a tough message to share. However, as Ms. Hill spoke with her personal and confident manner, we all felt ourselves strengthened to move forward and face this challenge as we always did during difficult times. Among the teachers and staff, there was a debate about when and how we should tell the children and parents. Should we wait until after the holiday? Should we wait for more definitive medical information? What kind of details should we tell the children? We had a lot of questions, but wanted to respect Ms. Hill's preferences. These were difficult waters to navigate.

I do not think I have ever felt so unprepared and powerless. To help me move forward, I hunted for articles and book chapters to find how others dealt with this type of challenge. I also consulted with my district supervisor and a few of the more seasoned school psychologists. Their insights varied and even conflicted regarding what they would do if they were in this type of situation.

I also needed to stay within professional bounds of what was expected of me. Ultimately, Ms. Hill and Mrs. Clark decided to write a letter to the parents. The letter, in a sealed envelope, would go home with students on the following Monday afternoon after everyone returned from Thanksgiving vacation. We decided to hold another short meeting with the staff and teachers before students arrived back to school on Tuesday morning. Rather than announcing on the intercom about Ms. Hill's illness, teachers were given a note to read with their classrooms on Tuesday morning. Ms. Hill also wanted to visit with each classroom and answer questions. Even facing such a serious illness, she was centered on helping her students and school community. That morning and for the rest of the week, we also had extra counselors and school psychologists at the school to offer students, teachers, and staff supportive counseling as needed.

Living with Uncertainty

The weeks that followed felt like an emotional rollercoaster ride. Everyone was excited to hear even the slightest bit of encouraging news. Before Christmas, Ms. Hill reaffirmed that she was fighting this illness and weathering her chemotherapy better than was expected. Although she lost her hair and looked pale and weak, her smile was the same. She talked about her cancer as if it were an uninvited guest that rudely and unexpectedly squatted in her body. It was as if she was standing toe-to-toe with her unwelcome visitor, demanding that the cancer leave.

We were on edge once again in mid-January when she was hospitalized. From what we could gather, she had contracted a cold that quickly turned into pneumonia. She was placed in intensive care and was not allowed to have any visitors. However, within a few days, she phoned to let us know that she was doing much better and had "rounded a corner." She was a fighter.

Ms. Hill did not want the children and school staff to worry about her over the weekend. I remember letting the children know on Friday that things were looking much better for our principal. We all heaved a huge sigh of relief.

Facing the Reality of Death

However, late Sunday night I received a phone call from the school secretary. The message was short: "Ms. Hill died." Inside my head I assertively asked and answered my questions: She what? She died? She died! She was alright on Friday. This could not be true. How could she be dead?

I remember sobbing and hoping that all of this was a dream. But, it was real. I took a very deep breath and slowly exhaled. Then I went into action. I needed to have instructional handouts for parents and teachers and I needed to provide options for classroom and school-wide memorial activities that would help the children and staff get through this challenging time. Although the school counselor and I focused heavily on the children's needs, we also did not want to overlook the importance of keeping teachers and staff emotionally anchored. We offered an extra hour after school for teachers and staff who wanted to check in and talk about how they were feeling. It was good to see teachers talking with their colleagues, hugging, and pulling together. This unity was a key ingredient that strengthened our hope and our stamina across the coming weeks.

DESCRIPTION OF RESPONSE

Feeling Supported

On Monday morning, the school district crisis team came to assist and we had plenty of counselors and school psychologists to support teachers and their classrooms, to help with small groups of students, and to meet one-on-one with those students who were having a particularly hard time. The district crisis team helped full force for the first week and then tapered off the assistance during the second week. My school psychology supervisor was very experienced and offered several recommendations that helped me support students and teachers over the days and weeks that followed. He gave me some handouts to review and suggested I revise the handouts as needed. I had confidence in his ability because he had years of experience assisting the district's schools following student and teacher deaths. He helped me process the experience and was there when I had questions and concerns. His support was something I leaned on when I felt overwhelmed or when I questioned my ability to effectively manage the situation.

I also relied on the vice principal. Mrs. Clark was strong and together we supported each other's efforts. We set aside time after school to talk and debrief. She was determined to keep the school afloat. We definitely felt the empty spot in the school. By default, Mrs. Clark was expected to fill that empty spot.

Although we had many serious talks, Mrs. Clark and I took time to laugh at funny things that students said and did. I enjoyed her honesty. I appreciated that we had a close relationship. When she was sad, we cried. When I was sad, we cried. We were a team. That felt good.

Attending the Funeral

Some decisions could not wait a week or two until things settled down a bit. Many parents asked if their children could attend Ms. Hill's funeral, the Friday following her death. Prior to the funeral, the children made handmade sympathy cards to express how much they appreciated their principal and to let Ms. Hill's family know how much she was going to be missed. The cards were delivered to her family by the counselor and Mrs. Clark.

The funeral was held in a church near the school. Consistent with the family's wishes, the district gave permission for teachers and students to attend. In preparation, prior to attending, teachers talked with their students about the funeral, what a funeral looked like and what would happen during the service. Letters were sent home to explain the opportunity for students to attend. Parents signed letters giving permission for their children to ride a school bus (provided by the district) and to attend the funeral. Many parents attended alongside their children. We had a lot of community support. Those students who did not want to go to the funeral and those students whose parents did not write a consent letter stayed at the school. Substitute teachers, school counselors from across the district, and community mental health counselors were provided so that teachers and staff could attend the funeral. Children who stayed behind participated in making banners and cards for Ms. Hill's family. They also hung banners in the school to honor Ms. Hill's life of service. Additionally, the gym and cafeteria were open for playing games.

The funeral was very short and "child friendly." Although not required, most of the children were dressed in their Sunday best clothes. They looked so mature! During the funeral, the children sat quietly and listened to the family's stories about Ms. Hill's life and accomplishments. Children were given the option of walking by the principal's casket, looking into the casket to see her face, and touching the casket. The funeral was a time to fondly remember Ms. Hill. The children enjoyed looking at the many beautiful flowers. Pictures of Ms. Hill were placed on a long thin table near the casket. The children enjoyed seeing pictures of Ms. Hill. Yes, our grandmotherly principal was once a child!

RECOVERY

Remembering and Memorializing the Principal's Life

Finding ways to remember and memorialize our loved one's life is one of the tasks of grief as outlined by Worden (2008) and Wolfelt (2002). This task was especially important to Shoshone Elementary survivers because remembering Ms. Hill was their way of valuing and respecting her many contributions to the school community. Although *remembering* was at times hard for us, we learned to cope with our sadness while holding on to our memories. Over time, our intense sadness was softened by our happy memories and the support we felt from one another.

Stained glass window. The entry to the school was brightly lit with natural light. A wall of square window panes stretched from a few feet above the floor and extended to the ceiling. My idea was to create the effect of a large stained glass window in the entry. I bought several rolls of clear contact paper and cut numerous windowpane-sized pieces for each classroom. I also bought a wide variety of colored tissue paper so that children could stick small pieces of tissue paper onto the sticky contact paper. When completed, students held their creation up to the light. The effect was similar to a "stained glass window." I then gathered the students' stained glass windows and taped the pieces together on the large window panes in the school's entry. Everyone loved the effect of this beautiful stained glass window and how it reminded them of Ms. Hill. We left the stained glass window up for the remainder of the school year.

Decorated concrete stepping stones. Ms. Hill was a gardener. She loved plants. Several years ago she created an herb and flower garden in a protected area alongside the school. To honor Ms. Hill, several students created stepping stones to decorate the garden. In the fresh concrete, they created colorful designs with glass marbles, buttons, and pieces of shimmering tile. Now denoted as the "memory garden," the Parent-Teacher Association purchased a beautiful shiny silver ball on a pedestal. In memory of Ms. Hill, the silver ball was engraved with her name.

Sensitivity to students' memories of Ms. Hill. There were times that I temporarily forgot, but reminders everywhere brought back to the reality of Ms. Hill's death. Several months after the funeral, I was conducting a classroom group. We discussed how individuals with disabilities can rise above their challenges, building on their strengths to develop amazing talents. I told them the story of Andrea Bocelli, a famous opera singer who was blind. Then, I played the song, *Time to Say Goodbye*, a duet by Sarah Brightman and Andrea Bocelli. One little girl in the front row began crying and needed to be comforted. Her tears reminded me that we all felt unprepared to say "goodbye" to Ms. Hill. We then discussed how separations are often sad times, but we do not want to forget those who are no longer with us. What was initially intended to be a lesson on rising above personal challenges turned into a support group on understanding our grief.

LESSONS LEARNED AND REFLECTIONS

Looking Back

Looking back on this experience, I think the biggest thing we did to help the students through this challenging situation was to "be there" for them, to allow them a safe place to talk about death and dying. The school counselor's door was always open for students, teachers, and staff who needed time to talk and express and process their feelings. Teachers were encouraged to talk about students' feelings and to alert the counselor if students needed individual time with the counselor, or me (the school psychologist) to visit classrooms and process feelings. Experts in grief emphasize the importance of providing opportunities for children to talk about death and to feel supported as they ask questions and express their feelings (Balk, Zaengle, & Corr, 2011; NASP School Safety and Crisis Response Committee, 2015; Wolfelt, 2002; Worden,

2008). Openness to discussing death and grief is an important strategy to teach parents and teachers, offering children a large safety net of support.

As I look back at Ms. Hill's death and how we as a school and community adapted and lived with such a huge loss, I am surprised by our resilience and how we all continued to come to school each day! I was amazed at how strong and supportive the teachers were as they continued to teach, maintain high academic expectations, and gently support each of their students.

Although initially difficult to face, over time I was comforted that each time I entered the school, I saw reminders of our dear principal. Her portrait was in the hallway outside of the office alongside a row of pictures documenting the elementary school's history. Her herb and flower garden remained beautiful and cheerful. Our "stained glass" window in the entry generated a warm, happy, and spiritual feeling as we walked into the school each day. The office area was brightened by flowers that students and teachers routinely replenished.

None of us would ever forget Ms. Hill. None of us wanted to forget.

REFERENCES

Balk, D.E., Zaengle, D., & Corr, C.A. (2011). Strengthening grief support for adolescents coping with a peer's death. *School Psychology International, 32,* 144–162.

NASP School Safety and Crisis Response Committee. (2015). *Addressing grief: Tips for teachers and administrators.* Bethesda, MD: National Association of School Psychologists. Retrieved from: www.nasponline.org/resources-and-publications/resources/school-safety-and-crisis/addressing-grief/addressing-grief-tips-for-teachers-and-administrators

Wolfelt, A.D. (2002). Children's grief. In S.E. Brock, P.J. Lazarus, & S.R. Jimerson (Eds.), *Best practices in school crisis prevention and intervention* (pp. 653–671). Bethesda, MD: National Association of School Psychologists

Worden, J.W. (2008). *Grief counseling and grief therapy: A handbook for the mental health practitioner* (4th ed.). New York: Springer.

Section 4

School Crisis Response to Death by Suicide

INTRODUCTION

Death by suicide among school age youth is a tragic event that can be prevented if caregivers and friends know the warning signs and how to respond. Youth at risk usually do not seek help, but often show warning signs to friends, classmates, parents, or trusted school staff. Warning signs should never be ignored (Erbacher, Singer & Poland, 2015). School personnel have a legal and ethical responsibility to respond to suicidal intent and should know the risk factors, protective factors and warning signs. Schools should have clear response procedures and trained mental health professionals and teams.

Suicide can have a profound impact on a school community. The school's response to intense, complex student emotions requires a balance of supporting the grieving process while preventing imitative thoughts and intent (Hart, 2012).

Prevalence

Suicide is the second leading cause of death among 10–19 year olds (CDC, 2013; Heron, 2016), and 17.7% of high school students report serious consideration of suicide, 14.6% had a plan, and 8.6% made an attempt, (CDC, 2016). While males account for about 84% of deaths by suicide, more females contemplate and attempt suicide. There are mediating factors, with rising rates of risk and attempts among those experiencing mental illness, depression and hopelessness; targets of bullying; lesbian, gay, bisexual, transgender, and questioning (LGBTQ) youth, especially in unsupportive environments (Hatzenbuehler, 2011); impulsivity and attention-deficit disorder (Sheftall et al., 2016); cultural and ethnic minorities; and military veterans (CDC, 2016).

Risk Factors

Situational crises and stressful events can increase suicide risk, especially when concurrent with chronic predisposing factors like depression or substance abuse, but no single factor is highly predictive (Miller, 2011). It is emphasized that risk factors can be addressed and treated.

Summary of suicide risk factors:

- Psychological Disorders—depression, bi-polar, impulsivity, alcohol/drug abuse.
- Previous suicide attempts.
- Family Factors—child maltreatment, parenting or family problems, economic hardship, death of a parent, parental divorce, family history of suicide attempts.
- Bullying and Cyberbullying—frequent bullying has been associated with depression, hopelessness and suicidal behavior (Gould & Kramer, 2011).
- Sexual Orientation—LGBTQ youth are at significantly higher risk for bullying and suicidal behavior than their straight peers (Eklund & Gueldner, 2012).
- Culture and Ethnicity—minority groups such as Native Americans and Latinos coping with acculturative stress have increased risk (Eklund & Gueldner, 2012).
- Self-Injury—self-injurious behavior such as "cutting" alone may not indicate suicidal intent, but indicates increased risk (Whitlock et al., 2012).
- Situational Factors—Access to firearms or other lethal means; Situational crises include breakup of a close relationship, trouble with school or police authorities, deportation or incarceration of a parent, academic failure, death of a loved one or anniversary of death, unwanted pregnancy or abortion, humiliation in front of peers, financial hardship, social isolation or rejection, serious injury or illness, increased pressure at school or home, community violence, increased home caregiver responsibilities, history of trauma or abuse (Lieberman et al., 2014).
- Contagion—The contagion effect increases risk that the suicide of a peer, family member, or other person will be imitated; Most at risk are those in physical or emotional proximity, such as witnesses or close friends, and those who are depressed or having suicidal ideation.
- Glorification of suicide by the media can contribute to contagion and clusters.

Protective Factors

Protective or resiliency factors can mitigate risk factors, decreasing risk of suicidal behavior (Eisenberg & Resnick, 2006; Gutierrez & Osman, 2008; Sharaf et al., 2009). When a child or adolescent is considered at risk, it is suggested that school, family and friends build protective factors around the youth.

Summary of suicide protective factors:

- Family warmth, cohesion, open communication and stability.
- Peer support, positive social networks, and a sense of belonging.
- School and community connectedness, safe and bully-free environment.
- Positive school climate and sufficient school-based mental health providers.
- Cultural or religious beliefs that discourage suicide and affirm life.
- Coping and problem-solving skills, frustration tolerance and conflict resolution.
- Interpersonal competence and academic or vocational success.
- Good self-esteem, resilience, and sense of purpose and satisfaction with life.
- Areas of interest and/or talent such as music, sports, arts.
- Easy, stigma-free access to medical and mental health resources.
- Suicide prevention programs, awareness of resources and crisis hotlines.

Warning Signs of Suicide

Warning signs are observable behaviors that can indicate suicidal thinking or intent. They can be considered "cries for help" or "red flags" to be taken seriously, and never to be ignored or kept secret. They signal the need to directly inquire about suicidal thoughts. If such thinking exists, there must be appropriate interventions.

Summary of suicide warning signs:

- Suicide threats—direct or indirect, verbal or written statements like "I want to kill myself", "The world would be better off without me" or loss of meaning in life.
- Plan or method or access to lethal means is disclosed, or lethal means is sought.
- Suicide note, plan, or online posting—describing a clear method and access to lethal means signals imminent danger.
- Previous suicidal behavior or attempts—powerful predictor of future behavior.
- Making final arrangements—writing a will or giving away valued possessions.
- Preoccupation with death and suicidal themes—excessive talking, drawing, reading and/or writing, music, movies about death may suggest suicidal thinking.
- Referencing own death or funeral, joking about it.
- Changes in behavior, appearance, hygiene, friends, thoughts, and/or feelings—depression, especially with hopelessness; sudden happiness, especially when preceded by depression; increased social isolation;

withdrawal from friends, family, society; changes in eating or sleeping habits; loss of energy, or fatigue; increased impulsiveness; reduced interest in previously important activities.

- "Masked" depression or emotional distress—acts of aggression, alcohol or substance abuse, sex, and/or risky behavior.
- Self-injury such as cutting, scratching, or burning the body.
- Inability to concentrate or think clearly, adversely affecting academics, school attendance, work, or other activities.
- Feeling humiliated—school discipline problems, threat of incarceration, recent psychiatric hospitalization.
- Searching the internet for methods or watching documentaries about suicide.
- Disinterest in future plans—"I won't be here anyway".

Sources: Adapted from Erbacher et al.,
2015; Lieberman et al., 2008

PREVENTION AND PREPAREDNESS

Goals of suicide prevention include identifying and referring at risk students for effective treatment, and generally reducing risk factors while enhancing protective factors. Since suicide rarely happens without warning signs, students, teachers, and parents/guardians may be in the best position to recognize the need for help. Warning signs must never be ignored. Intent never kept secret. Knowledge and action can save lives.

School Climate and Role in Preventing Suicide

A foundation of suicide prevention efforts is a safe, nurturing school climate where there are trusting relationships and connectedness among students and adults, and high levels of student satisfaction. Universal prevention programs are most effective within "positive school environments characterized by interpersonal warmth, equity, cooperation, and open communication (Miller, 2011)."

School counselors, psychologists, and crisis response teams must be prepared to intervene and assess when a student is identified at possible risk for suicide. Teachers, administrators, staff and parents must have training and consultation in risk factors, protective factors, warning signs, prevention, and procedures for effective response. School staff must be especially vigilant for students who are vulnerable because of individual or situational circumstances. Prevention programs must be developmentally and culturally responsive, with planning that includes materials in primary languages, interpreters, contacts with community agencies, and understanding the rituals, traditions, protective factors, and stressors of diverse school populations (Lieberman et al., 2014).

Reducing Stigma

Suicide has been stigmatized, contributing to the myth that talking about it puts suicidal ideas in people's heads. This belief is false and destructive, as it can obstruct prevention programs and support for survivors after a suicide. It is important to de-stigmatize discussion of suicide, and seeking help for depression, substance abuse, and other mental health problems. Schools can educate about mental health, open communication about depression and suicide, and provide access to school-based resources. Schools can reinforce the idea that getting help for a problem is not a sign of weakness, but rather an act of honesty and courage.

Safety Planning and Risk Assessment

Schools should review their information, resources, and sources of consultation on suicide and other crises (Reeves et al., 2010). It is suggested that districts or schools designate a trained suicide prevention expert, who can receive and act upon reports from teachers and others concerned about a student (Erbacher et al., 2015). However, response and risk assessment should be a collaborative, team effort.

While a school safety team develops prevention strategies and programs, a trained response team is prepared to intervene when there is a crisis. Members of both teams can overlap and comprise a trained risk assessment team led by school mental health providers such as school psychologists, counselors, and social workers. When a student at possible risk for suicide is reported to the risk assessment team, they conduct a comprehensive assessment. Schools can respond to suicidal crises as needed, or take a proactive approach with screening programs. *Preventing Suicide: A Toolkit for High Schools* (SAMHSA, 2012) recommends universal screening to identify students at risk for mental health concerns and provide for further assessment and referral if needed.

Evidence-based Prevention Programs and Resources

Signs of Suicide (SOS)—an example of an effective program for adolescents is *Signs of Suicide*. SOS encourages discussion and educates students and school staff about the warning signs of depression and suicide, and how to intervene. It also provides a short questionnaire that screens students for depression and suicide risk.

Anti-bullying programs—since bullying and suicidal behaviors share some common risk factors, bullying prevention programs can address both peer- and self-directed violence. Anti-bullying toolkits have been developed by the Alberti Center (http://gse.buffalo.edu/alberticenter) and The Empowerment Initiative (http://empowerment.unl.edu). Youth who frequently bully others and who are frequently bullied are at increased suicide risk, with LGBTQ youth at especially high risk. Programs that encourage acceptance of diversity are the Gay–Straight Alliance and Born This Way Foundation (www.bornthiswayfoundation.org).

Gatekeeper training—school gatekeepers such as students, teachers, administrators, staff, and parents can be taught about suicide, risk factors, warning signs, and how to get help for self and others. Gatekeeping emphasizes trust development by school staff with students, so they feel comfortable referring themselves or peers for possible suicidal behavior. The trained gatekeeper has knowledge and skills to identify and engage with youth at risk, and to connect them with caring school or professional resources as needed (Erbacher et al., 2015; Walsh et al., 2013).

Social media gatekeepers—students and educators can be trained as gatekeepers to identify suicide risk in social media. A staff "social media manager" with others, can monitor networks for high risk students, provide safe messaging about prevention, resiliency, community mental health resources and suicide hotlines, and educate about cyberbullying, shared prevention roles, positive use of social media, kindness apps and supportive communities (www.cyberbullying.us; www.LittleMonsters.com). Students can be encouraged to report cyberbullying, and to use supportive language on social media. Parents can be encouraged to get involved in their children's social media and work with Facebook and other sites to remove disturbing messages or images. Information about depression and suicide can be placed on school district websites (Erbacher et al., 2015).

Suicide prevention lifelines—inform about lifelines, including the National Suicide Prevention Lifeline: 1–800–273-TALK (8255), text START to 741–741, or suicidepreventionlifeline.org.

Means restriction—Since firearms are the most prevalent and lethal means of death by suicide, they and other means such as weapons, medications, ropes, etc. should never be accessible to youth and especially those at risk, without constant supervision.

Other resources and organizations:

- Preventing Suicide: A Toolkit for High Schools (SAMHSA, 2012)
- After a Suicide: A Toolkit for Schools (AFSP & SPRC, 2011)
- Local community and cultural mental health agencies
- American Foundation for Suicide Prevention (www.afsp.org)
- Suicide Prevention Resource Center (www.sprc.org)
- Centers for Disease Control and Prevention (www.cdc.gov)
- National Association of School Psychologists (www.nasponline.org)
- American Association of Suicidology (www.suicidology.org)

- The Jason Foundation (www.jasonfoundation.com)
- The Trevor Project (www.thetrevorproject.org)

INTERVENTION

Brent and colleagues (2013) emphasize treatment *early* in the course of identification and hospitalization. Interventions should be coordinated among school and community treatment providers, and immediately target risk factors such as depression and substance abuse. Interventions can concurrently strengthen protective factors like parent support and monitoring, a life-affirming social network, and coping and problem-solving skills.

Intervening with a Potentially Suicidal Student

When there are clear warning signs, the student should immediately be asked whether he or she has suicidal thoughts. Failure to ask directly may not provide the permission to share such thoughts. If there are suicidal thoughts, intervention must be immediate. No-suicide contracts cannot be depended upon to keep the student safe, and may provide a false sense of security. Instead, a *safety plan* must be developed that involves adult caregivers in suicide-proofing the environment, and providing supervision and treatment.

Suggestions for response by teachers, school staff, and parents:

- *Talk* with the student about suicide without fear of discussing the subject—encourage asking for help and connect with caring adults for ongoing support.
- *Know* risk factors, warning signs, and referral procedures.
- *Remain calm*, showing empathy and not distress that can block open discussion.
- *Listen* non-judgmentally to the expression of thoughts and feelings to understand the emotions that led to considering suicide.
- *Avoid minimizing* the student's emotional distress with statements like "You should get over it" or "Everyone has problems".
- *Supervise constantly*, staying with the student until transfer to a caregiver such as a crisis response/risk assessment team member.
- *Ask directly* if there is a suicide plan, and if necessary, safely remove the means.
- *Express concern* and reassure that help is available—the student can feel better in the future.
- *Respond* immediately, escorting the student to a designated school crisis/risk assessment team member, administrator, or mental health provider.
- *Join* the risk assessment, sharing detailed information about the student's thoughts and behavior that indicate risk.

Source: Adapted from Lieberman et al., 2008

Suggestions for friends and peers to help prevent suicide:

- *Know the warning signs*—take programs that teach risk factors and warning signs.
- *Talk with friends* without fear, listening to thoughts and feelings, letting them know you care, but not trying to prevent suicide without adult help.
- *Never make a deal* to keep a friend or peer's suicidal thoughts secret.
- *Tell a responsible adult* such as your parent, your school counselor or school psychologist—don't delay, thinking your message will not be taken seriously! Even if you are not sure of suicidal intent, find a trusted adult who will listen.
- *Act against bullying* and for positive connections among peers and adults.
- *Ask if there is a crisis response team* to help and if there is none advocate for one.
- If anyone is having suicidal thoughts, in addition to telling a trusted adult, you or the person at risk can call 1-800-273-TALK (8255) or text START to 741-741 or connect online to suicidepreventionlifeline. org.

Suggestions for school crisis response/risk assessment team members:

- After being escorted to a response team member—usually a school psychologist, counselor, or social worker, and until a collaborative risk assessment generates a plan—constantly accompany the student everywhere, enlisting other team members to monitor while private phone contacts are made.
- Collaboration among school administrator and response team for ongoing consultation and support while making difficult decisions is useful and reassuring.
- Document parent contact, or if parent is unavailable or uncooperative, document contact with protective services, regardless of the assessed degree of risk.
- Involve responsible parents in a *safety plan*, providing them treatment referral if needed, other school and community mental health agencies as appropriate, and guidelines for constant monitoring and removal of all lethal means.
- Law enforcement can be called if a student resists, attempts to flee, or if parents are uncooperative in addressing risk—the team should establish a relationship with law enforcement, who can possibly seek temporary hospitalization.
- Follow up to provide resources and referral, if needed for the student and family.

Source: Adapted from Lieberman et al., 2008

Risk Assessment

Since a suicide risk assessment may be a student's first experience with a mental health intervention, a safe, caring environment should be provided. While the student provides most information, teachers and caregivers can inform about home and school behavior, recent stressors, and family mental health history (Eklund & Gueldner, 2012).

The initial risk assessment at school takes into account *risk factors, protective factors, warning signs*, pursues *suicide inquiry* with pointed questions, possibly *screening instruments* for emotional status, and *clinical judgment*. The assessment generally assigns the *risk level* as *low, moderate*, or *high*, and these levels determine interventions to reduce risk. Use a form to **document** assessment, rationales, interventions, follow-up, and all contacts.

The risk assessment team determines whether the student is safe to return to general routine or whether there is risk for lethal harm to self or others. If risk of suicide is unclear or high, intervention can be stressful for the youth and family. The goal is to keep the student safe and develop a comprehensive care plan. Youth at risk are generally referred for an evaluation by trained professionals, to determine if psychiatric hospitalization is needed (Eklund & Gueldner, 2012).

Suggested questions to ask student at risk during suicide inquiry:

- Are there thoughts about suicide? Have you ever thought about suicide or harming yourself? Is the idea of suicide acceptable to you?
- Were there previous attempts? Have you ever tried to kill or harm yourself?
- Is there a plan? Do you have a current plan to kill or harm yourself? (High Risk—more specific plan).
- If there is a method student plans to use, is there access to means? (High Risk).
- What are student's main stressors, problems or reasons to die?
- What are student's reasons to live? Are there helping resources to support coping? (Positive answers may reduce risk).
- What are student's current feelings? Any signs of depression or desperation?
- What is student's level of hopelessness or perceived burdensomeness?
- Explore—Is there evidence of mental disorder such as hallucinations, delusions or commanding voices?

Sources: Adapted from Lieberman et al., 2008;
Lieberman et al., 2014

Suggested questions to ask parents, teachers, and staff:

- What warning signs initiated the referral?
- Has the student demonstrated sudden or dramatic changes in behavior?

- What support system surrounds the student? (feeling isolated/alone increases risk).
- Is there a history or current episode of mental disorder? (depression, alcohol or substance abuse, conduct or anxiety disorder, or co-morbid problems).
- Is there a family history of suicide attempts or death by suicide?
- What is the student's demonstrated level of impulsivity?
- Is there a history of recent deaths or losses, trauma, bullying or victimization?
- What are the student's current problems and stressors at home and at school?

Sources: Adapted from Lieberman et al., 2008;
Lieberman et al., 2014

Risk Level and suggested interventions (**DOCUMENT** all actions):

- Low Risk (moderate risk factors/ideation/no plan or intent/strong protective factors)—supervise and reassure student; advise parent; help connect with school and community outpatient resources; suicide-proof environments; mobilize support system; develop a *safety plan* that includes treatment as needed, identifies circle of caring adults and peers, promotes help seeking when needed, promotes communication and coping skills; provide prevention hotlines, websites, and app.
- Moderate Risk (multiple risk factors/ideation with plan/no intent/previous behaviors/few protective factors)—apply "Low Risk" interventions and consider "High Risk" interventions such as admission for evaluation if needed.
- High Risk (psychiatric disorders or severe situational crisis/plan and intent/access to lethal means)—supervise student at all times; hand off ONLY to either a parent/guardian who clearly commits to seek immediate admission for mental health evaluation, or law enforcement, or a mobile crisis responder for transport to evaluation; prepare a reentry safety plan for return from mental health hospitalization, including a meeting where student, parents, school, and community mental health representatives make appropriate follow-up plans.

Source: Adapted from Lieberman et al., 2014

Engaging Parents/Guardians

When a student is at risk for suicide, the parent or guardian must be contacted and informed of the threat, usually by a school psychologist or school counselor representing the risk assessment team. They should be asked about the availability of weapons, pills, or other lethal means and the need to secure such items. Even if the student denies suicidal intent, when information infers a threat, the caregivers must be notified and enlisted in prevention (Eklund & Gueldner, 2012; Lieberman et al., 2014).

The school psychologist or counsellor should try to arrange an immediate "face to face" meeting with the parents/guardians. This meeting is crucial to get assessment information, warn caregivers of the threat, educate about protecting their child, assess their cooperation, determine insurance coverage, and provide any needed referral information. The student may resist parental involvement, but can be helped to understand that in most cases, the caregiver is needed to improve the situation. Prior to bringing the family together for a discussion, it may be helpful to brief the caregivers about the seriousness of the problem and discuss constructive ways they can respond to their child (Lieberman et al., 2014).

Some parents readily understand the history of concerns for their child, while others may deny or minimize the threat. If parents are cooperative, a signed release of information should be obtained and immediate referral made for further evaluation as needed. If parents are unavailable, uncooperative, or there is basis to infer that the child may be abused or at increased risk at home, law enforcement, mobile crisis, and/or child protective services should be contacted to make certain there is transport for further evaluation as needed. When referring to the most appropriate community agencies for support, cultural, developmental, and sexual orientation issues should be considered. Follow up to be certain caregivers followed through with referrals for evaluation or treatment, and document every step in the assessment and intervention process (Lieberman et al., 2014).

Reentry from Hospitalization

The transition from hospital to school requires a careful, collaborative approach with the student, family, school-based mental health, school staff, and outpatient mental health services. All parties must meet and make an effort to involve the student in developing a reentry safety plan to support the ability to comfortably resume normal routine. Monitor peers and social media to prevent bullying and victimization. Create a network of social support by adults and peers in school and at home. Modify academic program to not overwhelm the returning student (Eklund & Geuldner, 2012; Lieberman et al., 2014).

POSTVENTION

Suicide in a school community requires a careful response that supports the opportunity to grieve, but does not glorify or romanticize the act. The postvention response must assess impact on the school community, identify and support affected students and staff, prevent contagion or risk for imitation, address stigma, and inform the community. *After a Suicide: A Toolkit for Schools* (AFSP & SPRC, 2011) is a valuable resource.

Suicide Postvention Process and Procedures:

- **Confirm the death and the facts**—confirm the death with authorities and family, and if possible, whether it was a suicide, before labeling the cause of death.
- **Contact the grieving family**—offer sympathy in person, asking how the school can help, offering resources, protecting personal effects for the family, identifying friends of the deceased, and discussing wishes for public or private funeral.
- **Assess level of response needed**—underestimating the suicide impact can lead to insufficient resources for assistance, while overestimating the impact can result in extensive, unnecessary services that may sensationalize the death. School administration and response team leaders should consider the nature and consequences of the event such as whether it occurred on campus, how well known the student was, emotional (close friends, family) and physical (witnesses) proximity of survivors, whether many students knew intent, and recent traumatic incidents such as a previous suicide, when anticipating extent of need.
- **Mobilize crisis response team**—based upon assessment of response needed, notify members of the school team and if necessary, the district level team. To assure sufficient response level, some district team members can be "on call" if needed. Response team should immediately meet for initial briefing and planning. News of the death may disseminate quickly by social media and cell phones, necessitating immediate intervention with the school community.
- **Provide accurate information**—as soon as possible notify and meet with staff, encourage expression of feelings, determine *what* and *how* to tell students, coordinate plan to carefully inform students such as a classroom script that avoids unnecessary details, expresses sympathy without glorifying the victim or the act, and shares how to get support. Provide counseling, in class support, and substitutes for teachers if needed. Avoid public address announcements or school-wide assemblies to share news. Inform other district schools with siblings of the deceased or students who may be affected. Prepare letters sent home to parents/guardians, informing about the death, what is being done to support students, warning signs of suicidal thinking, and available family resources.
- **Triage**—implement evaluation procedures to identify students at risk for severe reactions, including those having close relationships with the deceased, those in proximity to the suicide such as witnesses, and those considered at risk due to mental illness, previous suicidal behavior, or lack of family or social support. Students closest to the victim may struggle with feelings of guilt, anger, sadness, rejection, or isolation. Encourage affected students to seek assistance. The risk assessment team should

be prepared to screen for those at risk. Psychological triage is crucial to match appropriate interventions with degree of need.

- **Interventions**—establish a range of interventions according to individual need, at the "universal" level to support all students, "targeted" level for those needing more attention, and "intensive" level for those needing therapeutic treatment. Facilitate social support systems and *safe rooms* that encourage parents, students, teachers, and staff to refer self or others for support. Upset students should be encouraged to stay at school and use supervised safe rooms to express their emotions and get help. Individual, small group, classroom, and caregiver interventions should be available for those in need, but without imposing unnecessary interventions on those coping effectively. Educate that it is normal to express a variety of emotions, but be alert for misdirected blame or anger that could result in self-harm or violence toward others. Use teachable moments to educate about grief and provide opportunities for constructive action such as suicide prevention programs, learning about depression, and participating in life-affirming activities (Erbacher et al., 2015; Lieberman et al., 2014; Roth, 2015).
- **Reduce suicide contagion**—contagion occurs when suicidal behavior is imitated, and rarely, can result in a cluster of suicides in a school or geographic area. Postvention strategies to minimize contagion include: avoid sensationalism and unnecessary attention to the suicide act, not glorifying or vilifying the victim, identify and support those at risk, encourage referring or seeking help when needed, plan with community caregivers, enhance protective factors, and emphasize universal prevention strategies and *safe messaging*. Acknowledge the memory of the deceased while clearly and unambiguously encouraging students to distance themselves from the irreversible self-destructive act (Erbacher et al., 2015; Hart, 2012; Lieberman et al., 2014; Roth, 2015; Zenere, 2009).
- **Address social stigma**—stigma surrounding suicide must not hinder honest, sensitive discussion, prevention efforts, support for survivors, and encouragement to seek help for emotional problems.
- **Work with the media**—since the way the death is reported can have a profound effect on the community, a school district spokesperson should communicate with the news media, emphasizing that graphic, sensationalized or romanticized descriptions of suicide can lead to contagion and must be avoided (Hart, 2012). If there is an article, avoid suicide as a front page story, avoid photos of the victim or placing the word "suicide" in the caption, but provide helpful information such as warning signs, mental health referral options, and suicide prevention hotlines. A written statement with helpful information can be prepared for the media.
- **Monitor social media**—a small group of students and supervising faculty can monitor social media and social networks, performing a gatekeeper role that identifies possible suicidal language and risk. The group can also

provide *safe messaging* like "Suicide can be prevented" and "Available resources include . . ."

- **Memorials and funerals**—Use caution when memorializing a suicide, using best practices to avoid contagion. A brief, respectful expression of sympathy on campus such as a moment of silence without glorifying the victim or the act is suggested. Avoid formal ceremonies or permanent memorials such as yearbook dedications, tree plantings or plaques that honor the deceased. Options might be participation in a suicide prevention program or mental health organization. Spontaneous memorials that do not sensationalize the act should remain until after the funeral or about 5 days. School should not be cancelled for the funeral, but if the grieving family opts for a public funeral, students may voluntarily attend with their parents/guardians permission and support. Monitor the funeral to identify students who appear depressed, withdrawn, or express suicidal ideation. School can provide resources before and after the funeral, for processing feelings. Remove the student's desk after the funeral or after about 5 days so it does not become an informal memorial (Hart, 2012; Lieberman et al., 2014; Roth, 2015).

- **Team debriefing**—after postvention the crisis response team must have an opportunity to debrief in order to process personal reactions, express feelings, plan for stress management and support, and learn from what worked or needed improvement for future responses. Care for caregivers is vital for team wellbeing.

- **Long term follow-up**—Erbacher and colleagues (2015) point out that "sudden deaths such as suicide often take much longer to process due to the overwhelming shock and disbelief . . . along with difficulty comprehending why this happened." Debilitating *complicated grief* may manifest because suicide involves complex "social, emotional, and cultural issues including stigma, shame, and embarrassment" (Erbacher et al., 2015). As disengagement from crisis response is happening, it is crucial to emphasize the need for tertiary triage, which identifies vulnerable individuals who need monitoring and possibly treatment that could extend for weeks, months, or years. Anniversary dates of suicide can be a difficult time for affected students. Since the risk of contagion continues, school teams should be prepared to comfort those who are struggling and monitor those at risk, without glorifying the suicide. Consider positive anniversary activities such as providing suicide awareness and prevention materials or discussion (Erbacher et al., 2015; Hart, 2012; Jellinek & Okoli, 2012; Roth, 2015; Zenere, 2009).

REFERENCES

American Foundation for Suicide Prevention and Suicide Prevention Resource Center (AFSP & SPRC). (2011). *After a Suicide: A Toolkit for Schools*. Newton, MA: Education Development Center.

Brent, D.A., McMakin, D.L., Kennard, B.D., Goldstein, T.R., Mayes, T.L., & Douaihy, A.B. (2013). Protecting adolescents from self-harm: A critical review of intervention studies. *Journal of the American Academy of Child and Adolescent Psychiatry, 52*, 1260–1271.

Centers for Disease Control and Prevention (CDC). (2013). Mental health surveillance among children—United States, 2005–2011. *Morbidity and Mortality Weekly Report, 60*(2), 1–35.

Centers for Disease Control and Prevention (CDC). (2016). Youth risk behavior surveillance—United States, 2015. *Morbidity and Mortality Weekly Report: Surveillance Summaries, 65*(6), 1–174. Atlanta, GA: Center for Surveillance, Epidemiology, and Laboratory Services, U.S. Department of Health and Human Services.

Ecklund, K., & Gueldner, B. (2012). Suicidal thoughts and behaviors: Suicide intervention. In S.E. Brock & S.R. Jimerson (Eds.) *Best practices in school crisis prevention and intervention* (pp. 503–523; 2nd ed.). Bethesda, MD: National Association of School Psychologists.

Erbacher, T.A., Singer, J.B., & Poland, S. (2015) *Suicide in schools: A practitioner's guide to multi-level prevention, assessment, intervention and postvention*. New York: Routledge.

Eisenberg, M.E., & Resnick, M.D. (2006). Suicidality among gay, lesbian, and bisexual youth: The role of protective factors. *Journal of Adolescent Health, 39*, 662–668.

Gould, M. & Kramer, R.A. (2011). Youth suicide prevention. *Suicide and Life-threatening Behavior, 31*, 6–31.

Gutierrez, P.M., & Osman, A. (2008). *Adolescent suicide: An integrated approach to the assessment of risk and protective factors*. DeKalb, IL: Northern Illinois University Press.

Hart, S.R. (2012). Student suicide: Suicide postvention. In S.E. Brock & S.R. Jimerson (Eds.) *Best practices in school crisis prevention and intervention* (pp. 525–547; 2nd ed.). Bethesda, MD: National Association of School Psychologists.

Hatzenbuehler, M.L. (2011). The social environment and suicide attempts in lesbian, gay, and bisexual youth. *Pediatrics, 127*, 896–903.

Heron, M. (2016). Deaths: Leading causes for 2014. *National Vital Statistics Reports, 65*(5).

Jellinek, M.S. & Okoli, U.D. (2012). When a student dies: Organizing the school's response. *Child and Adolescent Psychiatric Clinics of North America, 21*, 57–67.

Lieberman, R., Poland, S., & Cassel, R. (2008). Best practices in suicide intervention. In A. Thomas & J. Grimes (Eds.), *Best practices in school psychology V* (pp.1457–1473). Bethesda, MD: National Association of School Psychologists.

Lieberman, R., Poland, S., & Kornfeld, C. (2014). Best practices in suicide prevention and intervention. In P. Harrison & A. Thomas (Eds.), *Best practices in school psychology: Systems-level services* (pp. 273–288). Bethesda, MD: National Association of School Psychologists.

Miller, D.N. (2011). *Child and adolescent suicidal behavior: School-based prevention, assessment, and intervention*. New York: Guilford Press.

Reeves, M., Kanan, L. & Plog, A. (2010). *Comprehensive planning for safe learning environments: A school professional's guide to integrating physical and psychological safety—prevention through recovery*. New York: Routledge.

Roth, J.C. (2015). *School crisis response: Reflections of a team leader*. Wilmington, DE: Hickory Run Press.

Sharaf, A.Y., Thompson, E.A., & Walsh, E. (2009). Protective effects of self-esteem and family support on suicide risk behaviors among at-risk adolescents. *Journal of Child and Adolescent Psychiatric Nursing, 22,* 160–168.

Sheftall, A.H., Asti, L., Horowitz, L.M., Felts, A., Fontanella, C.A., Campo, J.V., & Bridge, J.A. (2016). Suicide in elementary school-aged children and early adolescents. *Pediatrics, 138* (4).

Substance Abuse and Mental Health Services Administration (SAMHSA). (2012). *Preventing suicide: A toolkit for high schools.* (HHS Publication No. SMA-12-4669). Rockville, MD: Author.

Walsh, E., Hooven, C., & Kronick, B. (2013). School-wide staff and faculty training in suicide risk awareness: Successes and challenges. *Journal of Child and Adolescent Psychiatric Nursing, 26*(1), 53–61.

Whitlock, J., Muehlenkamp, J., Eckinrode, J., Purington, A., Baral Abrams, G., Berreira, P., & Kress, V. (2012). Nonsuicidal self-injury as a gateway to suicide in young adults. *Journal of Adolescent Health, 52,* 486–492.

Zenere, F.J. (2009). Suicide clusters and contagion. *Principal Leadership, 12,* 1–5.

8 Suicide Postvention Using the PREP<u>a</u>RE Model

Terri A. Erbacher

INTRODUCTION

Hearing about the death of a student is never easy. Hearing that the student chose to take his own life is even more difficult to grasp. Regardless of our levels of training and experience, none of us want to face this situation. Are we ever prepared for such a tragedy? How do we continue our work after a student suicide? How do we begin to help the other students—and the staff— work through such devastation? Many questions surface, many of which we will never be able to answer. In sharing one story of a suicide loss, I venture to share how our school response team handled such questions and the harsh realities we faced. I hope you never lose a student to suicide, but also hope that reading about our journey gives you another ounce of preparation, if you need it.

Hearing the News

I was sitting in a restaurant eating dinner with a close friend. We were chatting, laughing, and enjoying one another's company. My cell phone rang, but was in my purse so I didn't hear it. As I have a role responding to critical incidents, I frequently check my phone. I remember pulling the phone out of my purse, glancing at it and seeing a missed call from my high school principal. There was also a text saying "Please call me NOW." I returned the call and could tell immediately that there was panic in her voice.

"There has been a suicide."

My thoughts immediately raced to the many students with whom I've worked who have significant mental health issues. *Was it Samantha? Was it Johnny? Bobby? Who was it?* I wanted to know if this was a student I had been counseling, or had evaluated or gotten to know in some other way. Many of our students are the children of staff members, so it could be one of them. It all happened so fast, yet this moment of waiting for a name seemed an eternity. Principal Benny told me it was a student whom I had never met. His name was Conor,* a recent graduate. *I was thankful that I did not know him—I would be able to fully function in my crisis response role.* While I have served as a crisis responder for fifteen years and am a trainer of the National Association of School Psychologists' PREP<u>a</u>RE model (Brock et al., 2016), I was aware that, if I knew the student well and his suicide impacted me directly, I would not have been able to assist in coordinating the crisis response. I would be dealing with my own emotional reactions. I was glad I could move forward and help.

Principal Benny was already in a conference room with the other school administrators. They had all been on campus for a basketball game that night. Principal Benny described getting the notification on her cell phone and in the few moments it took to leave her bleacher seat, walk down the aisle, and exit the gymnasium, she observed changing expressions in other attendees. The news was spreading like wildfire on social media. She could see each person's jaw drop and smile fade. The gymnasium momentarily became silent; then whispers and hushed conversations began. The game continued as the timer counted down the minutes. People began to leave the gymnasium, perhaps to check on their own loved ones or to call friends to share the tragic news. The whole atmosphere had changed.

While I had Principal Benny and the other administrators on the phone, we began to discuss logistics. My first question: where was Conor's recent ex-girlfriend, Grace? *Was she yet aware? How would she take this news? Would she be at risk for suicide as a result? What about Conor's friends? What first steps should we take?*

Background

Conor was a 19 year old college freshman who graduated from our school the previous year. He and his long-term girlfriend had just broken up. He lived in a very close community of Irish Catholic families, whose children had all attended grade school together. Conor was an Eagle Scout, president of the National Honor Society, involved in sports, did volunteer work, and worked as a lifeguard. He had been honored with an award voted on by his senior class, which indicated the love his school and peers felt for him. Then, he went off to college. While he did not go far, he was no longer a part of this close community as he had been for 18 solid years. Conor apparently felt lost. He was an outstanding student attending his first choice college, but he no longer appeared to care about his academic success. Later, reports indicated that his first semester college grades were not stellar and not up to his potential. It was also discovered that his friends had been worried about him for a few months, not knowing how to help him. They communicated their worries to Conor's parents, who had also been concerned and immediately sought mental health treatment for Conor.

PREPAREDNESS

The Role of Social Media

During the last fifteen plus years that I have been responding to school crises, much has changed. In my early days, we had *time* to plan a response. When hearing of a student death, our crisis team could meet the following day to strategize. With the advent of social media, we must act immediately as students often discover information about these incidents well before we do. Within minutes, information is often shared with thousands of others as evidenced in our gymnasium. This brings about a multitude of challenges, including how to engage in rumor control. Since reports on social media are not always factual, we must get the facts out expeditiously, before rumors get out of control. Research suggests that schools be proactive in using social media (Erbacher, Singer & Poland, 2015). Consistently engaging parents and the community via social media with positive items such as sports events, school-based trainings, and honoring members of the school community gets parents and stakeholders habituated to checking the school's social media accounts. When a critical incident occurs, these stakeholders are accustomed to the sites being utilized and it becomes more natural to turn to them for information. It has become important that our crisis team members seek accurate information from police, the medical examiner, or the victim's family in to ensure accurate facts are posted.

DESCRIPTION OF RESPONSE

Contacting the Grieving Parents

Conor's parents were contacted by a school staff member who knew the family well—this is best, rather than a stranger calling a suffering family. Conor's mother answered the phone and was asked if she was comfortable with the facts of Conor's death being shared. She was initially hesitant. There is still significant stigma associated with suicide. However, our staff member explained that by sharing the truth, Conor's story might be used to break barriers, rid the stigma related to help-seeking behaviors, and hopefully begin to reach students we would not have known needed help. Letting students know they do not need to suffer alone might save another child's life. After this conversation, Conor's parents agreed that they did not want another parent to grieve as they did. They wanted to do anything possible to help prevent a future suicide. Thus, we were able to share the facts about Conor's death.

Reporting the Suicide

Importantly, we did not share any details about the method used in his suicide or where it occurred. We knew that providing details or graphic images can lead to contagion as others may try to emulate it in a copy-cat suicide. Contagion is the process by which one suicide may contribute to another. Adolescents are particularly vulnerable as they may identify with the behavior and qualities of a peer who has died (AFSP & SPRC, 2011). It is also important that we don't minimize the cause of the suicide. I recalled another local incident which the media portrayed as a death by suicide due to withdrawal of privileges by parents after a student received a bad grade in school. This oversimplification can frighten others, such as parents, who become fearful of appropriate discipline. It is important to remember that suicide and the emotionality associated with this choice are quite complex and there are often a multitude of factors present prior to a student's death. Further, approximately 90% of those who die by suicide have an underlying mental health condition (AFSP & SPRC, 2011). We did not yet know the depth of Conor's pain, but were careful not to oversimplify his suicide.

An important member of the crisis team is the public information officer (PIO), whether a school district superintendent, principal, or other designated media representative. It is suggested that the media representative be selected prior to critical incidents in order to be prepared with accurate information about suicide postvention and suggested suicide reporting practices. Best practices for reporting on suicide can be found at http://reportingonsuicide.org.

Reducing the Risk of Contagion through Social Media

It is critical that reporting be done appropriately to reduce the risk of contagion. For example, statements should include warning signs so readers know what to look for if they are concerned about a loved one. Reporting should also include available resources such as where to get help locally, and include the National Crisis Textline (Text HELP to 741741) as well as National Suicide Prevention Lifeline website and phone number 1–800–273-TALK (8255). The website www.suicidepreventionlifeline.org now includes an online chat function available for users who would prefer not to make phone calls. Many of today's youth rarely use a telephone, but are more familiar with text messaging as a primary means of communication. It is also strongly encouraged that schools put the warning signs and resources for finding help on their own social media venues. With Conor's death, information was included on all school social media sites as well as the main webpage, with online links where users could seek further information or resources.

Many social media sites have taken action to ensure users have ways to communicate should they be concerned about someone being suicidal. For example, social media sites, including Facebook, have simple ways to report these concerns immediately online. Directions on how to report concerns can be found at www.save.org/files/1/New_Documents/FBVideo_Instructions_ Link.pdf. Among the options for readers are contacting local emergency services, sending a message directly to the person of concern, or messaging a mutual friend who lives nearby. The reader can also report a concerning post or comment to Facebook administrators, who will send a pop-up message on the user's screen asking if the person would like help. Should assistance be accepted, the user is provided with a menu of options. This gives social media users an outlet *to do something* to help those they are concerned about—rather than feel powerless.

In addition to being proactive with a school or district's use of social media and posting the facts as soon as possible for rumor control, it is suggested that schools monitor social media for follow-up concerns. An administrator or other tech savvy staff person should be designated for this role. Since contagion is a significant concern after a suicide loss, monitoring of social media is of utmost importance. Students will sometimes post concerning thoughts, such as "I want to join you" after losing someone close to them through suicide. After Conor's suicide, students created an online Rest in Peace page. This was a public page viewable by school administrators, who continually monitored it, along with others, for comments by Conor's friends and classmates. In this case, many students rallied around one another for support, but thankfully, no concerning comments indicating further suicidal thoughts or behaviors were noted.

First Steps

Returning to our first steps and immediate response, Principal Benny and I briefly chatted about roles and responsibilities of our crisis response team. She would serve as incident commander and public information officer. In the terms our team uses, I would serve as the crisis response team leader and also coordinate student care. We began discussing logistics and triage, knowing we immediately wanted to find Conor's recent ex-girlfriend, Grace. We quickly learned that she was on an overnight retreat. While much of the community had already learned of Conor's death, Grace was not yet aware. We contacted the retreat leader's emergency number and found that the youths were engaged in activities that did not permit cell phones. We suggested to the retreat leader that these activities be extended until I arrived. I headed directly to the retreat center, which was just 15 minutes from my home. As I got in the car, Principal Benny called Grace's mother to meet me there. Knowing that how one learns about a tragedy can impact the traumatizing potential, I wanted to ensure Grace was told as gently as possible. I also wanted her greatest support to be there—she is very close with her mother. It felt strange to realize how thankful I was to be able to tell Grace in person so that she would not find out about Conor's tragic death via social media. It is rare these days to have this luxury.

There were so many other logistics to attend to . . . Prior to getting in my car to go to Grace's retreat, Principal Benny and I made a quick game plan until we could chat later in more depth. The building's two school counselors were called to the campus and were immediately available for any student who wanted to be seen at that moment. Many students were on campus for the basketball game. A *safe room* was identified for this purpose and the school also opened their sanctuary/meditation room as a quiet space for anyone who needed it. Students gathered there to quietly sit together and reflect. With many in shock, the room was very somber and still, offering comfort. With beginning logistics taken care of, administration had tasks to complete while I headed for the retreat center.

Sharing the Information

Grace's mother met me at the retreat center as planned. We met in the parking lot. This was our first time meeting in person, though we had spoken on the phone previously. The retreat leader pulled Grace from her session once we arrived. The moment she saw me and her mother sitting there together, her face dropped. She knew something horrible must have happened and quickly guessed it was about Conor. Grace dropped to her knees in shock. She stared blankly for a few seconds, repeating the word "no." The tears began to flow shortly thereafter. I sat with Grace and her mother for nearly an hour. We barely scratched the surface of her feelings. Finally, Grace indicated she had run out of tears, her distress was contained, and she wanted to go home. We encouraged her to stay off of social media as rumors were running rampant and would not help her effectively process what had just happened. She agreed to try and indicated the following day that while she did not stay off social media completely, as is typical for teenagers, she drastically limited her use.

When Grace headed home, the retreat staff and I made a plan to tell the rest of the students. Due to social media, some attendees were becoming aware as soon as their sessions ended. Others had not hurried to get to their phones and did not yet know. We sat all attendees together in the PREPaRE model's *psychoeducation* format. As opposed to a *classroom meeting*, which is designed primarily to disseminate facts, student psychoeducation groups are a bit more in depth, providing strategies students can use to help themselves and their peers cope with the crisis (Brock, 2011). As we began to discuss the tragedy, dispel rumors, and start to ask students for suggestions on how they've coped with hard times in the past, we realized that some students were greatly impacted and their phones were already ringing off the hook. Others did not know Conor and were not significantly distressed by the news. We therefore began the process of triage (discussed in more depth later), and broke the retreat attendees into two separate groups that were more homogeneous in the extent they were affected. We then continued to discuss coping strategies in these groups.

Between my assigned tasks at the retreat center, I was consulting by phone with the school principal in order to determine the appropriate language to use in the letter that would be sent to parents and posted on the website/social media for stakeholders and the community. As noted, we wanted to make

verified facts regarding Conor's death available as quickly as possible. And once everyone was informed . . . now what?

Reestablishing Social Supports

At this point, the harmony and peacefulness of the retreat, even for those who did not know Conor, was difficult to maintain. Many of the students affected by Conor's death wanted reunification with caregivers. They called parents to be picked up early to return to the comfort of their families, homes and beds. For teens, peers are often their greatest naturally occurring support. For this reason, some students wanted to be available should Grace need them, while others wanted to go be with friends. As it was getting late at night, some parents agreed to pick up other children who lived nearby. As we were not comfortable releasing children to anyone other than a legal guardian without written permission, the retreat staff and I quickly created a procedure for this. We allowed guardians to grant permission via email. Due to our online system, we had email addresses on file to verify each parent's identity. We quickly created a spreadsheet for monitoring who was being picked up by whom, and parents began to arrive. Presented with the comfort of their parent at this challenging time, a few students who had kept a brave face crumbled tearfully in their parent's arms. Parents too were tearful, some for the loss of Conor and some just grateful to see their child alive and well. While this situation of students being at this retreat center was unique, similar reunification procedures would have been put into place had this occurred during a typical school day.

Triage

We continued the process of triaging the students who remained at the retreat center. Psychological triage is the process through which the level of intervention each student might need is identified (Brock, 2011). First, those with emotional proximity are those who had a close relationship with Conor and might include friends, family, neighbors, and teammates, etc. These students would need to be seen immediately. Second, those with physical proximity are those who witnessed the death or found the body. As these students were at the retreat center, we knew that none of them had physical proximity to Conor's death. But we needed to determine who might have emotional proximity. Many who were close to Conor had already gone home. There were five remaining students who described themselves as being close with Conor. Since it is important to have students process their loss in homogeneous groups, we had these five students in one group and the remaining ten students in another. While the ten students did not have a close relationship with Conor, they watched the tragic events unfold with Grace and her friends. They were also saddened to hear of a peer feeling so distressed that suicide was an option. They grieved for their classmates who knew Conor and wanted to process this as well.

A third component to triage is being aware of those students who have preexisting vulnerabilities. Perhaps a student has internal vulnerabilities such as a prior history of depression or suicidal thoughts and behaviors, or external vulnerabilities such as a recent stress or death of a friend or family member. These students would be in need of immediate attention in the aftermath of a crisis situation (Brock, 2011). My thoughts quickly went to one student who had previously lost not one, but *two* immediate family members to suicide. Due to these familial tragedies, she had seriously considered suicide in the past, to the point of making a plan and we had recommended psychiatric hospitalization. As we were obviously concerned about her wellbeing, we chose to tell her about Conor's suicide individually, before telling the remainder of the group. As soon as I was done meeting Grace, I sat with this vulnerable student for some time. I had met her previously, multiple times, so we had already built strong rapport. She took the news without much affect, indicating that she did not know Conor. I was fearful that her blank reaction might be denial, and did not want to leave her alone. As I met with her, Principal Benny contacted her guardian (her aunt) to take her home. I would follow up with her again in the coming days. As the crisis events unfolded, we absolutely could have used more mental health staff on this night and knew we would need additional support the next day.

Determining Level of Response

Once Conor's death was publicly announced, our high school had offers of support from other local high schools as well as from our county crisis team. Following the PREPaRE model, we realized that determining the appropriate level of response needed is essential (Reeves et al., 2011). Under-responding may mean students who need to be seen are not, whereas over-responding may give students perceptions that they are not safe, that the crisis situation is worse than it is, and that they do not have the resiliency, internal coping strategies, and naturally occurring supports to handle this crisis without significant outside support. We decided that minimal and building levels of support were not sufficient. With our two counselors, one social worker, and one school psychologist on staff, we did not feel we were equipped to handle the significant number of students who indicated the prior night that they would likely seek guidance. We decided district level of support would be perfect. As this was a nonpublic school, this entailed reaching out to our county crisis team.

The County Crisis Team

Our county crisis team officially began in 2013, utilizing the PREPaRE model of crisis response (Brock et al., 2016). I had become a PREPaRE trainer in

2008 when we started an agency crisis team. Our agency, called an Intermediate Unit, provides programs and services to school districts in our geographic area. PREPaRE training was successful and we continued to provide trainings throughout the county. With the horror of mass shootings like Columbine and Sandy Hook, our longer-term goal was to expand our team from just our agency to include every school district in our county. While it took time to build, we have expanded team capacity and increased sustainability. Rather than just a few people capable of responding, we now have nearly 30 trained responders willing to pitch in when needed. We've created response teams of 5–6 members, which allows more response flexibility and camaraderie within each team. We also established a PREPaRE Leadership Team—a core group of five members who are trained to mentor newer team members and to *lead* teams responding to critical incidents in schools.

With this team of dedicated individuals, I knew we could secure the support needed the following day. In collaboration with the county crisis team administrator, the school principal, and myself, we decided to have four additional counselors on site, with three supplementary crisis counselors on call should we need more support. Again, making this determination is an educated guess in an effort not to over or under respond. As the night was drawing to a close, I was already exhausted. I knew I would be up before the sun the following day and was grateful that additional staff would be on campus to assist. As it turned out, all four counselors were needed and utilized throughout the next two days. We did not need those on call, indicating we made an accurate estimation of our needs. Now that we had informed our student body about the facts of Conor's suicide, initiated triage, written a letter to inform the community, planned our *safe rooms* for students seeking help, and secured additional counseling, we considered what else we needed to do.

Self-Care

There is a reason I am not waiting until the end of this narrative to discuss self-care. Self-care should not be the last thing we think about as responders. Rather, it should be an ongoing process of considering our own needs in order to be more effective caregivers. As I was up late meeting with students and quickly realized the following day would also be stressful, I needed to take care of myself. Returning from the retreat center, it was too late to go for a run, my best form of relaxation and stress relief. I decided to do some light reading and make sure I got some good sleep! However, as soon as I got to my neighborhood, some friends who were mindful of my work started asking me questions about the case. Conor's suicide had already been highly publicized in the media, and everyone seemed to have questions about it. After such a rough evening, I did not want to discuss the case all night. I had to return refreshed in the morning. So I told neighbors that I was happy to sit on my deck together, but needed a break from talking about this tragic loss. My lovely neighbors sat with me, sharing lighthearted conversation and laughter. That was probably the best impromptu stress reliever I could have had. As I crawled into bed for a few hours of fitful sleep, the lists of things to consider for the following day continued to spin through my mind. I had to awaken early as our crisis team and administration were meeting at 6:30 a.m. to give us time to prepare before our meeting with teachers at 7:30am.

Teacher Meeting

To ensure our teachers were provided updated facts and prepared for typical reactions, including techniques for responding to student reactions, we asked them to arrive early for a *caregiver training* (Brock, 2011). As Conor's suicide greatly impacted our entire school community, every teacher was on time for our early morning meeting, except for two with child care details. While the meeting purpose was to share the facts, most teachers already knew them. We ensured that rumors were squashed and questions answered. Principal Benny let the teachers know the procedures for the day and provided them with a script to read to each of their homerooms via a *classroom meeting* format (Brock, 2011). Our assistant principal informed staff of the *safe room* locations and steps for referring students. At my urging, he further suggested that teachers let students go freely to safe rooms and not be gatekeepers. Rather, we asked that they allow our mental health staff to determine which students needed help in the safe room, and which could return to class. This relieved them of responsibility and allowed them to return to teaching for the students who were ready.

As the administrators spoke, I observed teachers' faces and they seemed sad and shocked. I thanked them all profusely for what they were about to do for our students—just by being the caring, kind support they always are. I wanted to ensure that they had some tools, such as sample responses to typical student questions after a suicide. This included reassuring students that they could not have predicted Conor's suicide and that it is okay to say "*I don't know*" when students ask *why* it happened (Erbacher et al., 2015). It is a complex task after a tragedy to figure out when to return to teaching. Some students are in need of structure and routine while others feel that returning to "normal" makes it seem like no one cares their peer died. We encouraged teachers to cancel tests and other high-stakes presentations, and advised that they keep lessons flexible for the day. We suggested they begin each class by recognizing the loss, sharing their own sorrow (to a reasonable degree), and letting their students know that we are all grieving together. I assured teachers that I would be available should they need to debrief, share concerns about students, or just vent stress. As we had additional support staff that day to meet with students in safe rooms, I was assigned tasks of continuing triage, facilitating procedures, collaborating with the principal on how to respond to the media, and being the support person for teachers and staff as needed.

Psychological Crisis Interventions

While our school uses safe room interventions, the PREPaRE model refers to the task we engaged in throughout the following day as a *classroom-based crisis intervention*. In addition to providing crisis facts, this intervention provides an opportunity for students to share their own story and reactions to the loss and to empower students (Brock, 2011). However, identifying space to conduct these interventions on the following day proved to be a challenging task. Most rooms in the building are utilized. We were able to secure space in each of the counselor's offices and in the teacher lounge, the meditation room, and academic department offices. Our procedure entailed each student first checking into the guidance suite where an administrative assistant tracked their names and the time of their visit, guided them to an available room and staff person, and monitored those being seen. While students were seen as expeditiously as possible, there were occasionally some in the waiting room. We were able to keep waiting time to a minimum. We kept in mind that additional staff were on call if needed. Many students fit into similarly affected homogeneous groups seen together, such as Grace's volleyball team and Conor's former baseball teammates.

We did not expect to see Grace in school so soon, though she did come in later in the day. She arrived with her mom and indicated that school was a safe haven for her, the place she wanted to be. I met with Grace one on one, providing *individual crisis intervention* (Brock, 2011), while her mother remained in my waiting room. As I began to prioritize her crisis problems, I recalled a case a few years ago, in which a 9th grade student's older brother had killed his long-term girlfriend. While working with his little sister, I caught myself assuming that her biggest concern would be her brother's current status in jail and that he may remain there for life. Prioritizing from her perspective, I could not have been more wrong. Her greatest problem at that point was actually intense grief she was experiencing at the loss of her brother's girlfriend, who had become like an older sister to her. I try to remember this any time I am in this phase of intervention. When we are assessing student priorities, it is easy to make false assumptions. As we talked, it became clear that Grace's greatest crisis problem was the guilt she felt for breaking up with Conor. Among many other reactions, guilt can be very common after a suicide loss (Erbacher et al., 2015).

Continuing Triage

Triage is an ongoing process that continued throughout the following day as students needing support were identified. Often, students came to the safe rooms independently. Other sources of referral were our school secretary, who knew Conor's neighbors, and our transportation director, who pointed out those Conor knew since elementary school. We worked our way through these lists, calling each student down to check in. Sadly, as Conor had texted several close friends his whereabouts just prior to his death, they had gone looking for him hoping to find him alive. A friend's father found Conor first— and too late. His friends in the vicinity, unfortunately witnessed the tragic scene. They would need significant support and possibly individualized therapeutic services outside of school as a follow up.

Many students I worked with in the past who witnessed a suicide scene experienced flashbacks, visualizing that scene over and over. It has been reported to me that flashbacks are most frequent at night. One student noted that after Conor's death, he kept extremely busy, but when he laid his head upon his pillow to sleep, visions of what he had witnessed flooded his mind. Sleeping became a challenge. He had trouble getting these flashbacks out of his head. As we continued to triage students with emotional and physical proximity to Conor and his death, we continued to seek out students with pre-existing vulnerabilities to ensure they were counseled.

Complicated Grief

It is hard to describe the grief that one feels upon hearing that a friend or loved one has taken his or her own life. Many of the students we saw that first day in safe rooms asked the question: "Why?" *Why did Conor take his life? Why was he so distraught that he thought suicide was his only option? Why did he feel life wasn't worth living?* And, even more challenging for these students was to integrate these questions: *Why didn't he tell me he was hurting? Why didn't I see the signs? Why didn't he come to me for help? Did I pay enough attention and could I have saved him?* Many of these questions remained unanswered as Conor's peers tried to navigate their grief. Denial seemed short-lived for most of them, including his girlfriend. Grace had a multitude of feelings that varied between sadness, confusion, anger, remorse, and significant feelings of guilt. She could not help but wonder if she was somehow the cause of Conor's death. Since they recently broke up, she wondered if she had pushed this already vulnerable teen toward wanting to end his life. She tortured herself over the details of their last interaction—*Had she been unkind to him? What were the last words she had spoken to him? Did he know that she still cared about him although they were no longer a couple? Was there anything she could have done to prevent his death?* While she was clear she needed space from him, she never wanted him harmed. These thoughts would continue to torment her for the weeks to come.

Many of Conor's friends had similar thoughts and questions. They gathered around each other and in safe rooms for support. In this close community, their parents also gathered together in the evenings. This helped ensure these teens were not alone in their grief. We knew that teens often self-medicate after a tragedy, utilizing drugs or alcohol trying to feel better—even if only for a moment. While there was likely some of this going on, many of these teens found solace in their community and the comfort of being with friends and family. The grief associated with suicide is often referred to as *complicated grief* as this type of death overwhelms students with shock, disbelief, and the question *why* this happened (Erbacher et al., 2015). There is also significant stigma associated with a suicide loss that makes the grief process more complex. Our crisis team would hear more about the stigma experienced by Conor's family in the days to come.

The Following Days

As the community sought return to a sense of normalcy, everything appeared a bit somber. Everyone was tentative, afraid to say the wrong thing should someone else be secretly contemplating suicide. It seemed that everyone was walking on eggshells. Our district crisis team continued to provide safe room coverage for two more days—after that, it was sufficient to rely on the counseling staff in the building. A few students continued to come down daily—three close friends of Conor, and Grace. Since Conor was an alumnus, we did not have to worry about things such as his empty seat in class or clearing out his locker. We noticed that certain staff members were significantly distraught over this death. Many wondered if they could have done something differently: *Should I have spent more time building rapport with Conor? Would he have shared his pain with me had I let him know I was there for him? Did I miss the warning signs that could have saved his life?* We offered them services through the Employee Assistance Program and let staff know they were welcome in our counseling center any time. A suicide loss, particularly of a young person, impacts a community at large. With suicide being the second leading cause of death for youth nationally (CDC, 2014), it is a frightening public health problem. Yet, lives can be saved if we can identify those at risk and get them help. As often happens, our school principal was quite distressed. This was her first year serving in this position and she was thrust into a tragedy that would quickly redefine her role in a way she never imagined.

Memorials

The question of how to memorialize Conor's death was perhaps the greatest struggle we encountered throughout the postvention response. Conor's friends wanted to make t-shirts in his honor, displaying an image of his face. They also wanted to sell memorial bracelets that contained the last words Conor had posted on social media—essentially his goodbye quote. Our crisis team was knowledgeable about the research regarding memorials and did not want to sensationalize a suicide death (Reeves et al., 2011). We gently encouraged Conor's friends not to make the t-shirts. We asked Grace's opinion, and she stated she did *not* want to see Conor's face plastered everywhere as she walked down the hallway. Grace realized this would not help her move forward in her grief. Because these were her friends and peers, she wanted to discuss her feelings with them personally.

We also knew that not recognizing the loss at all could make students resentful (AFSP & SPRC, 2011). We carefully considered our steps. We compromised with students by agreeing to allow a memorial in the meditation room consisting of pictures of Conor and a candle lit in his memory. We chose the meditation room, a private space where students could choose whether to enter. We believed this location was a better option than a main hallway or lobby that forced students and visitors to face Conor's photographs as traumatic reminders. Research suggests that memorials be removed after the funeral or after approximately five days (AFSP & SPRC, 2011). We agreed to keep the memorial for the week, at which point the students would deliver the pictures gathered to Conor's family. Living memorials such as a donation to a charity or research foundation, or the purchase of a suicide prevention program are recommended because these types of memorials help students find meaning in their loss. As students began to raise money in honor of Conor, the administration agreed to schedule a local speaker for the following year to share her story of depression—and recovery—hoping to educate students on mental health and wellness.

Dreadfully, students began showing up to school the following week wearing t-shirts with Conor's picture on the back and his last words on the front. This was particularly distressing for some of his close friends as an unexpected traumatic reminder. Grace reported she would finally find herself having a stress-free moment, and Conor's face on a shirt would flash in front of her. While she spoke to Conor's closest friends about the impact of the t-shirts, word had not reached the entire student body. Some of the students were given disciplinary infractions as they had been directed not to wear these shirts, infuriating them more. We were now dealing with anger on top of the shock and grief students were experiencing. My one regret throughout this postvention response is that I believe we could have communicated more effectively with the *entire* student body about the traumatizing potential of such memorabilia. We could have prevented the angst experienced by students who did not understand the reasons for our decisions. They felt we were sweeping Conor's death under the rug and we were unable to help them understand that we had their best interests at heart.

Our crisis team has since revisited this topic while evaluating our postvention response and we have made changes to our plan. As recommended by *After a Suicide: A Toolkit for Schools* (AFSP & SPRC, 2011), we now have a memorial policy that includes acceptable commemorative practices regardless of the manner of death. Administrators are expected to adhere to the policy when making decisions. Having a policy in place makes it more understandable for students who may feel it is unjust to ban t-shirts and certain other gestures.

RECOVERY

Long-term Support

As grief is an ongoing process, we provided long-term support to students with a grief group. We waited a few weeks after Conor's death before beginning the group to allow the students' most raw feelings to subside—an approach beneficial for this type of group (Erbacher et al., 2015). This allowed us to determine who was appropriate for the group. Those who continued to experience raw or complicated grief might do better with individual therapeutic support. It is also important that members of a support group be homogeneous regarding emotional and physical proximity. For example, we did not want mildly affected group members to experience vicarious trauma from a severely affected peer who witnessed the suicide. For this reason, Conor's close friends who saw his body and also Grace would not be part of the group. After meeting with students to ensure appropriateness and secure permission forms from parents, there were ten students who all cared deeply about Conor, but were outside his closest circle. This grief support group ran for nine weeks. It was planned for eight weeks, but the students asked for one more meeting to celebrate Conor's life. Through these nine weeks, I had the gift of watching these students move through their grief.

When the group began, many students identified significant confusion, questioning why Conor chose to die, questioning God, and questioning the meaning and purpose of life. Through the nine weeks, it was miraculous to watch these students move to a place of appreciating life, finding joy in simple things, and finding greater compassion for their peers. One student, Craig, told of a peer who seemed "nerdy and isolated." Craig reached out to this quiet classmate, befriended him, and was seeing a movie with him the following weekend. Craig talked about now understanding that one small gesture could make all the difference in the world to someone who might be struggling. These transformations truly were a gift to observe and while I had hoped these students would learn through the group, I think I learned even more. I learned that there can be positive growth even in the hardest times—*if* you are open to it. I learned just how much resiliency our young students have. And I learned the power of the human spirit—with all the tragedy and horror in the world, most people are kind and want nothing more than to connect with those around them.

The Anniversary

A few days before the first anniversary of Conor's suicide, we heard reports of a student we will call Sean, posting a statement on social media that led his friends to believe he was considering taking his life on the train tracks. Several years earlier, our larger community had suffered the tragic death of two girls who took their lives in a suicide pact on the same tracks. This suicide pact had left our community reeling and contagion was always on our minds. We were terrified. We did not want to lose another student. We wondered if Sean knew the two girls or was friendly with Conor. We are so thankful to the students who reported concerns about Sean. This allowed us to follow up immediately. Trained in suicide risk assessment, I quickly pulled Sean from class, built rapport with him, and assessed his risk. While Sean reported that he did not know Conor or the two girls, he was feeling depressed and was indeed having suicidal ideation. However, he stated that he would "never do it" as he knew his mom "would fall apart and couldn't do that to her." Sean indicated that he felt isolated, not realizing that those who reported the concerns considered themselves close friends who cared greatly about him. Sean's suicidal thoughts and behaviors were not imminent and he did not have a plan, but we were able to get him help.

Upon contacting Sean's parents, his father quickly arrived at school. He was in a bit of shock, having no idea his son was depressed. At home, Sean did not isolate himself, was jovial, at times silly, and joked frequently with his parents and younger brother. His father did not believe Sean might be suicidal. But Sean's dad also knew Conor through the community and quickly realized that Conor's father probably would not have believed it either. Therefore, he readily agreed to take Sean to an emergency appointment at one of our local counseling centers that evening. The center provides immediate appointments for our students in need. It is important to build relationships with local resources before they are needed in crisis situations (Brock, 2011; Reeves et al., 2011), and for this reason we developed an ongoing referral base. Having counseling centers readily available can alleviate the stress of students having to wait for hours to be seen in a hospital or psychiatric crisis center. Sean continued to have consistent follow-up appointments with a therapist through the remainder of the school year and began to feel his depression slowly lift. The students who reported Sean's concerning statement just might have saved his life.

Our community continues to be fearful of another death by suicide. We continue to provide training for teachers on warning signs and continue our own training for mental health staff on suicide risk assessment, safety planning, and monitoring suicidality. We have learned the importance of communication with our students and staff. In Pennsylvania, Act 71 recently went into effect, requiring all school staff in grades six through twelve to be trained in suicide warning signs and prevention, and encouraging training for students (Pennsylvania General Assembly, 2014). This professional development requirement will ensure that teachers and school staff are continually updated on new trends and research regarding suicide prevention and that all new teachers are aware of the warning signs—as they too just might save a life.

* All names, and some details, have been changed to protect the privacy of those involved.

REFERENCES

American Foundation for Suicide Prevention and Suicide Prevention Resource Center [AFSP & SPRC]. (2011). *After a suicide: A toolkit for schools.* Newton, MA: Education Development Center.

Brock, S.E. (2011). *PREPaRE Workshop 2: Crisis intervention and recovery: The roles of school-based mental health professionals* (2nd edition). Bethesda, MD: National Association of School Psychologists.

Brock, S.E., Nickerson, A.B., Reeves, M.A., Conolly, C.N, Jimerson, S.R., Pesce, R.C., & Lazzaro, B.R. (2016). *School crisis prevention and intervention: The PREPaRE model, second edition.* Bethesda, MD: National Association of School Psychologists.

Centers for Disease Control and Prevention [CDC]. (2014). National Center for Injury Prevention and Control. Web-based Injury Statistics Query and Reporting System (WISQARS, 2014). Available online from www.cdc.gov/injury/wisqars/index.html. Retrieved on July 10, 2016.

Erbacher, T.A., Singer, J.B., & Poland, S. (2015) *Suicide in schools: A practitioner's guide to multi-level prevention, assessment, intervention and postvention.* New York: Routledge.

Pennsylvania General Assembly (2014, June 26). *Act 71: Public school code of 1949— Youth suicide awareness and prevention and child exploitation awareness education. P.L. 779, No. 71.* Available online from www.legis.state.pa.us/cfdocs/Legis/LI/uconsCheck. cfm?txtType=HTM&yr=2014&sessInd=0&smthLwInd=0&act=71. Retrieved June 17, 2016.

Reeves, M.A., Nickerson, A.B., Conolly-Wilson, C.N., Susan, M.K., Lazzaro, B.R., Jimerson, S.R., & Pesce, R.C. (2011). *PREPaRE Workshop 1: Crisis prevention & preparedness: Comprehensive school safety planning* (2nd edition). National Association of School Psychologists, Bethesda, MD.

9   Student Suicide on Campus
Response and Resilience

Kathy Bobby

INTRODUCTION

Typical of my morning routine, I was answering email when I heard a lot of commotion in the hallway. I heard the words, "Is this a joke? Is this a theatrical stunt?" and then I heard someone scream. As I ran out of the room toward the chaos, I repeatedly heard the words "a student is on fire." When I arrived outside, I came upon a sight that my brain just could not comprehend. What I saw was not a student on fire, but only what I can describe as an ashen figure of a person as if drawn with pencil strokes and shading with no facial features. Several adults had formed a human barricade around this rooted figure and kept asking him, "What is your name?" Other adults were frantically walking around and everyone on the scene had a look of shock registered on their faces. My immediate reaction was to help, but I could not think of a single thing to do. Finally, I asked the administrator who had taken the lead what I could do to help, and he asked me to go into the buildings and make sure that the blinds were drawn on every classroom that overlooked the scene. I then went back into the building and did just that. Unfortunately, the brand new blinds that hung on the classroom windows were not opaque, but were a somewhat sheer material that did little to block the horrifying scene outside. I returned to the front office, and it occurred to me to call the district office to inform my supervisor about what had happened. I called the office and left the rambling voicemail of a school psychologist in shock. I then called the school counseling coordinator, and she told me that she already was on the way. She asked me about how many other school counselors and school psychologists we would need to handle the situation, and I told her that I really had no idea. Little did I know that we would spend the next several days with over 50 mental health professionals, including both district employees and outside agency workers, responding to the student suicide.

Kathy Roberts (in Bernard, Rittle & Roberts, 2011)[1]

I wrote the above personal account six months after the tragedy occurred. My colleagues and I had been asked by the National Association of School Psychologists (NASP) to write about how we used the PREPaRE model (Brock et al., 2009) to respond. As we began collaborating for the article, I felt the need to let down my defenses and to pour out my account of the event. I did not know at the time that this would be referred to by national crisis responders as a "9/11" event, that my friends/co-authors and I would present this story to local and national audiences in the following years, and that I would be asked to read this account over and over as an opening for the PREPaRE 2 (Brock, 2011) professional training module.

When I was contacted by the editor of the present book, inquiring if I was interested in submitting my story, I definitely debated whether or not I wanted to "put my head back there". It's been five and a half years since the event. Although the details are now somewhat fuzzy, especially the order and timing of what happened, the overall experience immediately became a part of me and had a profound impact on my life.

The following text is an attempt to share my personal story following this event. I will share my memories, thoughts, feelings, reactions, and reflections related to the event and my experience with crisis work since. I will share the good, the bad, and the ugly of crisis work and will highlight issues related to the support of crisis responders. I will also share lessons learned and hope that readers share this information with their districts to improve overall crisis response.

What Happened?

A 16 year old boy drove to school one morning in December of 2010 with a full gas can in his car. He parked in his assigned spot and when he got out of the car, proceeded to pour gas all over himself. He then lit himself on fire and ran towards the front of the school building. Unfortunately, he did this at the time when many students were either already in their classrooms or were late arriving to school. Consequently, there were hundreds of eyewitnesses to the event and several first responders who immediately tried to help put out the fire by using fire extinguishers and blankets. Local emergency crews were called and the boy was taken to the hospital and quickly transported to a burn center in another state where he died later that week.

DESCRIPTION OF RESPONSE

How Did the School Respond?

Fortunately, the district's school psychologists had been recently trained in the PREPaRE crisis intervention model (Brock et al., 2009). Shortly after arriving at the school, the team of school psychologists and the school counseling coordinator pulled the curriculum "off the shelf" and quickly operationalized the model. The article, titled "Utilizing the PREPaRE Model When Multiple Classrooms Witness a Traumatic Event" by Lisa J. Bernard, Carrie Rittle, and & Kathy Roberts, and published in the October 2011 NASP *Communique*, describes how the model was used to respond to this crisis.

How Did I Respond?

I would call myself a *second* responder to the incident. I was not a "hands on" *first responder*; however, I definitely witnessed the "aftermath" of the event. Initially, I attempted to help to stabilize the physical environment in order to keep students safe. After that, I reported what had unfolded to district crisis responders, as the "horrific scene" had already been removed by the time they arrived. I then participated in the planning and execution of the district's crisis response utilizing the PREPaRE crisis intervention model.

At the time I was very aware that I was a potential crisis victim due to what I had witnessed. Consequently, the best and most appropriate decision made was when my friends suggested that I offer logistical help with the crisis response rather than participate in counseling. We all understood that I had experienced the traumatic event alongside many other staff members, students, and parents who witnessed the event, and that it would not be professionally prudent for me to provide counseling to others. It was almost inherently understood that it was important for me to be part of the crisis response to empower me to "do something" to help the situation, but that my assistance would need to be limited in scope. This element was critical for me when I think back. It allowed me to begin my own "healing process" right from the start. I have frequently used this knowledge to guide other "ground level" individuals who have a "personal stake/vulnerability" in the crisis that they are responding to.

Memories from Those Days

One of the difficult tasks that I was assigned was to collect and sit with "hands on" first responders in order to prevent them from leaving prior to debriefing with law enforcement. They were mostly teachers, and a couple of students and parents who tried to put out the fire and stayed with the victim. One of the memories that I have was staring at the "ash" on a teacher's black clogs as I rubbed her shoulders while we waited for her turn to report on the event.

Another memory I have on the following day was of a student who was late to school and missed the student led prayer/memorial service conducted at the flagpole and the parking spot where the student burned. She was very upset as she had brought a flower to place at the flagpole. I told her that I would walk her out to the flagpole to place her flower; however, I was intercepted at the front door by an administrator alerting me that the press were waiting outside for the opportunity to take a picture of the "raw emotions" of those close to the tragedy. I assured the student that I would place her flower and I escorted her to class.

I also remember that at one point a live video of the event was posted on the internet and a live interview between a news reporter and a student eyewitness was posted on a news feed. Both were quickly removed later that day much to my relief and surprise.

I remember listening to crisis responders debriefing at the end of the first and second days following the event. A handful of responders reported that they felt like they were experiencing *vicarious trauma* after listening to the trauma stories of the eyewitnesses. It was most traumatizing for responders to listen to the detail of the "sensory reports" of the witnesses. Debriefing was a critical part of the process for these responders as it gave them a forum to process their own emotions in a supportive environment.

Lastly, my daughter was a student at the school, and I remember being frantic to know if she had been an eyewitness to the event. When I was finally able to locate her and catch her eye, she looked away with shock and pain in her eyes. Later when we were in the "safety" of our home, I learned that she had not witnessed the event although many of her friends had.

RECOVERY

Personal Reactions to the Events

Following the event, I was very cognizant of the potential crisis reactions that I might experience, especially since we had provided this information to parents, teachers, and students. One reaction that I experienced for a period of time was that I physically reacted when I saw fire either in person or on TV. My heart rate would increase and the hair would stand on the back of my neck. I found this reaction very surprising as I had never actually seen the boy on fire.

Another reaction I found myself experiencing and still experience to this day is associated with a smell on the day of the incident. There is a peculiar scent that I had noticed every time I entered one particular door of the school building and for some reason that smell is etched in my memory and it takes me back to that day. I'm sure that I must have smelled this scent prior to the incident, but now it is associated in my mind with the incident.

I definitely found myself having difficulty concentrating; however, this reaction didn't seem to set in until a couple of months following the event. I think that I was on hypervigilant mode initially following the event, but after enough time elapsed and I started to relax, I found it very difficult to concentrate. This lasted for a couple of months.

I found myself pulling away from close relationships and getting closer with fellow first and second responders. I remember thinking at the time that people just had no idea what we had experienced, and I found that I had very little patience when the subject came up in conversation with those who had not been a part of the event. I got annoyed and even disgusted with people as they discussed the event and even felt like they were being voyeuristic. I don't remember how long these feelings continued, but eventually, they subsided and stopped.

I also found myself "over-inclined" to assist with other district crises even when my team was not on call. Soon after the psychologists and counselors were trained in the PREPaRE model (Brock et al., 2009), a structure was put into place whereby four district crisis teams were created. I was named leader of Crisis Team 2. We were scheduled to be on call one week per month for response. During the rest of the 2011 school year and for the next several years, I volunteered to respond to crisis events even when my team was not on call. My district also frequently requested my presence at a crisis when I was not on call. As I reflect on the extent of my crisis work over those years, I realize that my "over-involvement" with helping other people through their difficult experiences was possibly based on my need to feel a "sense of control" in uncontrollable situations for others. Perhaps a "transference" experience for me? Regardless of the cause of my responder hypervigilance, I found my involvement personally therapeutic as I continued to subconsciously process my own traumatic experience.

Another related experience was my involvement as a PREPaRE trainer. A month after the event occurred, our district provided us with the opportunity to participate in the PREPaRE Trainer of Trainers (TOT) module. A handful of colleagues and I attended this training and became certified trainers of the PREPaRE crisis intervention model. Following the suicide, my district decided to provide regularly scheduled PREPaRE trainings for administrators, school counselors, paraprofessionals, school psychologists, and any other school-based employee deemed appropriate to attend the training. It was expected that all PREPaRE trainers would present at least two trainings per school year, including trainings in neighboring districts.

When we began these trainings my colleagues suggested that I read aloud my "first hand" account from the *Communique* article at the beginning of every training to capture the audience's attention. It definitely captured their attention; however, whether it was a good idea "psychologically" for me to repeatedly read my account, I was never quite sure. It went something like this: I read my account, getting choked up along the way. My colleague then introduced the training, giving me time to "pull myself together", and then I returned to the podium to begin the training. The rest of the time we "tag teamed", frequently using anecdotes from our crisis experiences as a way to show how to operationalize the model. This format became very effective as evidenced by our training evaluations. As the years went by, I found myself become less apt to get "choked up" as I read my account. I assume that my more emotionally regulated reactions became a good example of the potential positive outcomes of systematic desensitization. I thank my colleagues for their "therapeutic" presentation idea.

LESSONS LEARNED AND REFLECTIONS

Discretion in the Use of Lockdown

As I think back to this event, there are many lessons learned. The first is that students remained on lockdown much too long. Attempting to contain the situation and allow time to operationalize our plans for *triage*, we kept students on lockdown for hours in their classrooms. This extended time of isolation for students and teachers may have added distress to the situation.

Supporting Appropriate Teacher Response

Another lesson learned is that some teachers did not monitor their own reactions as they tried to make sense of the event. This was evident through classroom conversations reported by students following the event. Some of these teacher initiated conversations apparently included disturbing details of similar burning incidents. Attempting to stabilize teachers and address these behaviors, the crisis team met with the entire faculty to discuss the concept that students come to school with their "personal vulnerabilities" and that some conversation topics directly related to crisis details are not appropriate and can cause more harm. Individual personal vulnerabilities of staff were also discussed and they were informed that the Employee Assistance Program (EAP) would provide on-campus counseling services the following day for those who needed this support. EAP brochures explained available services if they wanted support outside of school or at a later date. It is interesting that no staff sought on-campus services the following day even though they were highly confidential. This could be attributed to wanting to continue the "business as usual" adult composure, not trusting the "confidentiality" of the services, or adults often utilizing pre-existing coping strategies and familiar networks. No data was ever collected in the months following the crisis whether or not faculty engaged in counseling support related to the event.

Dispelling Rumors through Classroom Meetings

The importance of dispelling rumors is another lesson quickly learned. As we conducted the *Classroom Meetings* prescribed by the PREPaRE model (Brock et al., 2009), we quickly realized all of the misinformation being discussed by students and teachers. One student believed that the deceased student wore too much cologne, was working on his car in the parking lot, got too close to the engine, and "caught on fire." This and other misinformation was quickly corrected, thereby reducing the potential for additional psychological harm due to inaccurate information.

Controlling Social Media

Controlling social media access was another lesson learned from this event. Some students took video of the event as it occurred. This video was uploaded to social media by one student and for a time was available for public viewing. In addition, a reporter had gained access to a student and conducted a phone interview describing her personal account of the tragedy. This too was available on social media for a time that day. At some point later in the day, both uploads were removed from social media by local law enforcement.

During the event, most students were texting their parents and friends from other schools. This instantaneous communication added to the level of crisis impact on the district as a whole. Rather than being isolated to the individual campus where the crisis occurred, panic spread throughout the district. Parents called in, alarmed, and showed up at school to take their children home. Inaccurate rumors spread like wildfire regarding the "facts" of the incident, and the privacy of friends, family, and first responders was forfeited.

Documenting Student Contacts

Documenting individual student support became a very important activity of the crisis team. Initially students were grouped homogeneously, based on their degree of exposure and relationship to the event. Entire classrooms who witnessed the event participated in PREPaRE prescribed therapeutic *Classroom Meetings*. Students who were "hands-on" first responders with the burning student were identified and met with individually. Students who had prior experience with suicide ideation or family/friend suicide were identified and met with individually. Close friends of the student and the family were grouped and given extra attention.

Data on all students seen individually and in small groups was collected and documented in a database. Students were rated either *red*, *yellow*, or *green* based on the counselors' rating of their degree of emotional distress. Students who were red or yellow received follow-up either through on-campus counseling services or were referred to outside agencies. Parent contact was critical in the determination of service delivery. The database was regularly updated as students moved from red, to yellow, to green, and was a helpful tool to ensure that no students "fell through the cracks". The database was also helpful in the following years. If a student began to falter either behaviorally or academically, the psychologist or counselor was able to quickly ascertain the student's exposure to the event and determine whether it was a potential contributing factor to their current struggle. The database also served as a "safeguard" for the district as a means to document all the work completed in the wake of a crisis if the district was later accused of negligence.

Ongoing Communication and Cooperation with the Media

Another lesson learned is the importance of proactive and continued communication with the media during and after a widespread crisis event. It is important to protect the public from inappropriate, excessive exposure to a crisis event—removal of the eyewitness video and the personal account interview from social media. It is also crucial to gain the media's cooperation and educate them about the potential negative consequences when information is shared with the public. Specifically, the local newspaper published an article six months after the event with medical information explaining the possible cause of the suicide. The media did not consider that the article was published during final exam week at the recovering school. Had this been anticipated, school officials and the family could have asked the paper to postpone the article's publication until the week after the exams so the students could have been home with their families to process the new information. Unfortunately, the article became a major distraction causing significant negative impact on many individuals that week.

Ongoing and Long Term Emotional Support

Ongoing support and evaluation were provided to move the campus toward "psychological closure". Trauma-focused counselors were made available on campus to assess students whose parents were particularly concerned with on-going crisis reactions. Continued referrals were made to outside agencies if deemed appropriate. A campus-wide survey was administered to students, parents, and staff four months after the event in an attempt to measure the "emotional wellness" of the campus. An "anxiety" in-service about the symptoms and causes of anxiety, and proactive interventions to prevent it was provided to faculty and staff. Suicide prevention programs were introduced to students in following years, and the district suicide protocol was updated based on a nationally recognized model.

Student Resilience and Empowerment

One of the most important lessons learned is how resilient children can be. On the day after the incident, before he died, students gathered in the courtyard for a school-wide group photo in a gesture of support. I was keenly aware as the anniversary of the tragedy approached, and thought that many of the students would want to plan an event to commemorate the deceased student. There were many conversations and the best guidance was from a national crisis responder, who suggested allowing students to take the lead.

One student suggested that students wanting to congregate could meet in the senior lounge at lunchtime. The administration endorsed this action and it turned out that very few students needed to respond in an expressive way. It appeared that the students as a whole wanted to move on and move forward. When the year came to an end, it was decided to include the student in the yearbook as he would have been a graduating senior.

A couple of years ago, I began wondering about the potential negative impact of this event on student test scores and absenteeism rates of faculty and students. I pulled the state school report cards for the year prior to the event and for three years afterward and found that both the average test scores and the absenteeism rates for both students and staff remained relatively consistent through the years, thus empirically supporting the hypothesis that people are indeed resilient.

Potential Issues Associated with Crisis Work

In the time since this crisis, my colleagues and I have responded to numerous incidents including accidental deaths, student murders, and parent and teacher suicides. We remain vigilant in the utilization of the PREPaRE model (Brock et al., 2009) for crisis intervention and believe that we have accomplished positive outcomes through our psychological triage and first aid. While we have seen positive outcomes with the students we have helped, we have identified some potentially negative outcomes for crisis responders that can arise depending on the level of district support.

When individual crises occur, crisis responders are often out of their buildings for a whole day or more. If coverage is not offered by the district, responders are regularly expected to continue to meet all job responsibilities in addition to their crisis work. This dual role and responsibility can readily lead to "burnout" for the crisis responder. Too much crisis work can also lead to "burnout". It is important that the crisis work is shared by teams of individuals and that responders are required to take a break from crisis work every couple of years or as needed.

It is also important that crisis teams are empowered by the district to be the "crisis experts" when mobilized. We found that sometimes administrators, trying to continue being the lead decision-maker without understanding crisis response, could compromise the operations of the PREPaRE model, and make potentially harmful decisions. Educating administrators to be knowledgeable about the concepts of crisis response and the need to empower responders in the face of a crisis is critical for an effective collaborative approach and successful outcomes.

Lastly, certified PREPaRE trainers should be compensated for trainings they conduct. In their training, trainers are educated on the logistics of conducting workshops and reasonable fee expectation. Districts requiring certified trainers to regularly conduct trainings within or outside their district without compensation or coverage can be psychologically detrimental to trainers. "Caring for the caregiver" is a component of the PREPaRE model. When trainers do not feel adequately supported by their districts, "caring for the caregiver" becomes an oxymoron.

Concluding Lessons and Personal Reflections

How does one conclude a personal reflection of a major event that forever changed one's life? I am not really sure what the appropriate format should be other than perhaps offering up some summary thoughts to this essay.

- First, when a significant event happens in your life and you use it to help others, you end up helping yourself in the process.
- Second, the more difficult situations that we face and overcome or are exposed to in life, the less we tend to "sweat the small stuff".
- Third, as the PREPaRE model curriculum states, *it's not if a crisis occurs, but when a crisis occurs.* We need to recognize this knowledge and adequately prepare to respond in an appropriate manner when the need arises.
- Fourth, crisis work is difficult work. It is our ethical duty and responsibility to make the "caring for our responders" as high a priority as the crisis work itself.
- Fifth, we need to be cognizant of the fact that the media and social media are pivotal influences on the successful outcomes of a crisis response.
- Sixth, we all need to be more self-aware that our words and reactions to events impact those around us.
- Last and most important, it is part of human nature to be resilient and to move toward homeostasis as quickly as possible. This innate need should be recognized and utilized as a natural tool when overcoming a crisis situation.

NOTE

1. Copyright 2011 by the National Association of School Psychologists. Bethesda, MD. Reprinted with permission of the publisher. www.nasponline.org.

REFERENCES

Bernard, L.J., Rittle, C., & Roberts, K. (2011). Utilizing the PREPaRE model when multiple classrooms witness a traumatic event. *Communique, 40* (2), 10.

Brock, S.E. (2011). *PREPaRE Workshop 2: Crisis intervention and recovery: The roles of school-based mental health professionals* (2nd edition). Bethesda, MD: National Association of School Psychologists.

Brock, S.E., Nickerson, A.B., Reeves, M.A., Jimerson, S.R., Lieberman, R., & Feinberg, T.A. (2009). *School crisis prevention and intervention: The PREPaRE model.* Bethesda, MD: National Association of School Psychologists.

10 Suicide on the School Campus

Contagion Magnified?

Lea Howell and Richard Lieberman

INTRODUCTION

I vividly recall an interaction with a colleague at my district's "back to school" social engagement prior to my first year as a full time school psychologist. The associate principal at the high school where I would be spending most of my assignment greeted me with a big grin. He exuded a friendly confidence as we went through introductions. It is imprinted in my memory that as he graciously excused himself to socialize with other colleagues he said, "Glad you are joining our team. We'll keep you busy!" I took this as a trustworthy admission that my work life was not going to be easy. But I also pondered, "How hard can this job be?" And reassured myself, "You can handle busy."

Approximately six months later our school endured the most trying crisis in our school district's recent history. After the crisis I began to recall the interaction with the associate principal and couldn't help but feel that his "We'll keep you busy!" remark was prophetic. I pictured that moment like an overdone movie scene in slow motion, warning the eager psychologist beginning her career that things were about to get real.

It had been an effort to get this position. I was an educator my entire career, but found myself in a surprising struggle to find full time employment as school psychologist. I had recently obtained my masters degree in school psychology after seven years as a kindergarten teacher. It was a tough job market, but I had worked hard and proven myself in part time positions. Then I was finally hired! I recall when the lead psychologist told me the good news. She was excited for me, but spoke with an almost apologetic tone, informing me that the majority of my time would be at one of the district's three large high schools. She knew my expertise was with elementary students, but thought I would do well at a high school. I was ecstatic. Sure, I was nervous about how I would win over unruly teens, but I felt a blind sense of confidence. When a former colleague heard where I would be working she gave me a coy look and a thick "good luck with that school" comment. After the warnings,

I started assuming this school had quite a reputation. Despite the cautionary congratulations, I held my head high. I was ready.

I found the high school a quality institution with a strong staff that provides students with the opportunity to enter the next phase of their lives with an armor of skills to help them be successful. But it also is a place where academic rigor can fuel insecurities and many students feel excessive pressure to succeed. There is a perfect storm of high expectations at school, pressure from home, and mental health issues. I realize now that even before the crisis many staff members were battling with burnout due to large caseloads and difficult expectations from students, parents, and the district. The increasing number of adolescents suffering from anxiety, depression, and other emotional problems was already wearing on staff and would be even more difficult after the crisis.

The Day of the Crisis

I imagine everyone present at the incident has their own unique narrative about their experience of what happened. Mine is a mix of moments that are remarkably vivid and others that are a complete blur, including parts of the crisis response process and postvention efforts. This personal account is written with the intention to share my story and also to impart the knowledge I've gained from the process of consultation that was key during postvention. I think most professionals who have been through a career-changing situation must feel gratitude for the support they had, but also reflect on the ways things could have been handled more effectively.

The day of the crisis was a Friday and I was already exhausted. The night before, I had been talked into going to an emotionally charged program that educates teens about drinking and driving. The administrator invited me to be there so I could assist if students became upset about the sensitive material. After a graphic video, she announced to the group of nearly 30 students that I would be leading a processing group that I was unprepared to facilitate. I got home late and had a hard time getting to sleep. I wondered if the same uneasiness was also preventing my students from sleeping. Viewing horrific images can have impact on kids and adults alike. I dragged myself out of bed the next morning thinking I could get through the last workday of the week and finally get some rest. But I have learned that most work days never go as planned.

The lunch bell sounded when suddenly several special education students who had bonded with me came bounding into my office with a sense of panic, all shouting confusing information. I recall Janet telling me that she witnessed a student jump from the top of one of the buildings into the quad, where many students gathered to socialize before lunch. Janet was worried he was dead. I called the special education department and when I told a secretary what I suspected, she seemed flustered by the horrifying information and implied I had not contacted the right place. I calmly explained that she needed to tell my boss what was happening and request more mental health support on campus. I texted the psychologists whose numbers I had and asked them to come to the school as soon as they could. I don't recollect how much later I went out to the quad, but the images are frozen in my mind even though my access was at the opposite end from where the body lay. While the school was in lockdown, I could not imagine how many panicked students wandered around campus, not believing what they witnessed. A student had jumped to his death in the middle of campus in the middle of lunch and it was witnessed by hundreds of students and staff.

I immediately reported to the cafeteria which logistically became our crisis room for students, and spent most of the afternoon with other psychologists, counselors, and mental health professionals called to assist students who had immediate counseling needs. We saw a wide range of reactions, from those who still appeared shocked and numb, to those crying and grieving, to those acting silly and saying stupid things. Many students were already reporting anger or immediately trying to find a reason, someone or something to blame for the victim's behavior. Suicide is complex, so is the grief that follows. There were *many* severely impacted students who left campus, but not yet identified. Psychological triage for this event was going to be huge. The body was eventually identified as a 15 year old male 10th grade student. The bizarre end to an already surreal day came when notes found in the dead student's pocket warned a friend not to do the same thing. We took names and numbers and just when I thought my day was over, I was asked to accompany two police officers to the home of the girl who had been warned, to do a risk assessment. There was fear of a suicide pact. Sometimes we absolutely just have to get in the car and go! Fortunately, the girl and her parents were home. Everyone was cooperative, and we assessed the student as low risk. We set a follow-up plan and I finally got a chance to rest, sensing our return to school would be an extraordinary day. I remember thinking graduate school did not cover this!

DESCRIPTION OF RESPONSE

Phase 1: The Immediate Response

I was not involved in the weekend planning for the response when school resumed. I am not sure whether not being involved can be attributed to the overwhelming, unexpected nature of this event or because I was so new to the position. I admit that I assumed that the more experienced district psychologists would know what to do. I now realize that the planning over the weekend must have been a confusing and daunting task for even the experienced professionals. Our district did not have a formal crisis team. I simply tried to relax and recover, reassuring myself that the lead psychologist would communicate the plan to me when we arrived at school after the weekend.

All our district psychologists were asked to come to the school if they could. There was a tremendous influx of people, including from neighboring districts. While it was impressive to see so many people willing to help, I was quickly aware that it was chaotic, and that perceived anxiety was affecting the students being referred. It was announced that students needing counseling should go to the library, which could accommodate many students and not interrupt the cafeteria schedule. There were sign in sheets, but unfortunately little else to keep track of student concerns. A psychologist brought a triage form she found in her files, but it was rejected because it seemed difficult to use and probably did not meet the scope of our needs. It was difficult to determine who should see which student, and how to group students or whether to see them individually. Some kids were willing to wait a long time to talk to a familiar face like their psychologist or counselor rather than a person they thought would be gone at the end of the day. At times it seemed like a battle to deal with students who seemed to be malingering in the library to get out of class. Some responders created lists of students they had seen and their assessment of those they were concerned about. I recall often becoming agitated when colleagues verbally shared concerns about students with me, expressing the need to keep them on my radar and assuming I would remember everything they told me. We were doing our best, but it certainly was challenging with so many referrals.

The spot where the student died remained an eerie site in the aftermath. A colleague recalls that it was cleaned up over the weekend and looked normal by the time we returned to school. I don't remember the first time I looked at the spot in the weeks after the crisis, but I sometimes still feel strange walking in the area. I find myself wondering where it exactly happened. In the initial aftermath there was some awareness that memorials, especially on campus, might be problematic, but I had no resources with explicit recommendations on how to deal with memorials. Not surprisingly, a spontaneous memorial began to grow in the general area of the quad, but items were quickly moved, troubling some students. In retrospect, perhaps we should have let it build for a few days and then take the flowers and handwritten notes to the family. An area was informally designated for peers to leave notes, but we eventually realized this also posed unforeseen difficulties. Some of these notes indicated that the author was struggling with the death or with personal problems. We wanted to contact these students but some did not leave enough information for us to identify them.

Amidst my processing of the state of my school, I had to decide what to do about the trip I had already paid for to the annual convention of the National Association of School Psychologists (NASP). I decided I had to go. I didn't want to leave the students and staff at such a vulnerable time, but it felt right to take some time to seek the guidance and support of my fellow school psychologists. I soon found that the information I received at the convention created valuable momentum for the next phase of the response.

Phase II: Days, Weeks Later

Overwhelming Referrals, Using Resources and Valued Consultation

I attended as many sessions as I could about crisis response at the national convention. My eyes filled with tears in one session as I heard about a well-organized response team who had been trained in the PREPaRE model (Brock et al., 2009). I struggled with feeling frustrated about why my district did not have that training or something similar. I recall a text exchange with the lead psychologist while I was still at the convention that claimed, "things are calming down." I think she was trying to be reassuring, but I was becoming more and more aware that this crisis was not quietly going away. However, I became inspired and invigorated by the information I was getting. After these sessions I connected with presenters who all suggested I contact a co-author of PREPaRE and suicide prevention expert in my local area. Perhaps most importantly, from connecting with colleagues at NASP, I began to feel I was not alone.

I became aware that we needed to think seriously about the potential for suicide contagion. It was right there in my most valuable resource prior to consultation, NASP's Best Practices in School Psychology (2008). It clearly stated in the postvention section that there was "considerable scientific evidence that a suicide in the community or exposure to the suicidality in others raises risk in vulnerable children and adolescents." But I had no idea how to begin the postvention phase with limited knowledge. I was fortunate that my colleagues at the convention connected me with Richard Lieberman immediately upon my return. He validated my concern that the "things are quieting down" hope was completely unrealistic. We needed to prepare to carefully identify the most affected students. Recent research estimates that, for every death by suicide, 147 people are exposed and, among those, more than six will have their lives changed forever.

I now had access to *After a Suicide: A Toolkit for Schools* (AFSP & SPRC, 2011), one of most helpful, practical guides created as a resource for school mental health personnel. At times when crisis responders find it difficult to think, the toolkit provides the necessary resources and directions. I wished I had access to the toolkit immediately after the incident. Even weeks later I found sections in the toolkit about how to help students cope and how to limit contagion particularly helpful.

My fears about ongoing trauma were realized one morning when I saw almost a hundred names of students on a referral list, but had no idea how each child wound up on this list! It was clear we needed to back the truck up. To conduct an adequate risk assessment, each student would have to be re-interviewed, which could present the problem of re-exposure. I consulted with Rich, and in collaboration with a few crisis team colleagues, we set about creating a Triage Interview Form based on the psychological triage factors enumerated in the PREPaRE model. We prioritized students into three groups.

First, we wanted to identify those we feared were the highest risk group—students who witnessed or were physically proximal to the event. Next, we wanted to find those who were emotionally proximal to the victim such as siblings, friends and classmates. Perhaps the most difficult at-risk students to find were those who may not have been on the quad, who did not know the victim, but for whom experiencing others' grief ignited their own history of depression, previous suicidal thoughts or traumatic loss. While the grouping of students was initially helpful, we quickly found that some students appeared in more than one group, and those in all three groups revealed high risk for suicidal behavior.

We began reassessing each referred student with sensitivity about re-exposure. While students sometimes had to repeat information such as their physical and emotional proximity to the victim, we could now apply information about potential risk factors to identify vulnerable students. For example, I began meeting frequently with a student I had been assessing for special education services. He was emotionally at risk because he was friends with the victim and was on the quad when the tragic event happened.

At staff meetings we focused on their critical role both in crisis intervention and suicide prevention. Guidelines for staff included:

- Know that *how* you respond to the crisis will have an impact on your students.
- Let us know if you need classroom support, time to talk, or want to connect with counseling through Employee Assistance Program.
- Dispel rumors with facts, but avoid frightening details.
- Be flexible with your academic expectations in the classroom for a while, offering students an opportunity to talk about their crisis reactions.
- Validate/normalize a wide range of reactions and emotions.
- Offer helpful ways for kids to cope (handouts provided).
- Don't be afraid to talk with your students about suicide. You will NOT be putting ideas in their heads.
- Know and be prepared to recognize the *risk factors* and *warning signs* (handout).
- Know the referral process and if concerned about a student, escort or arrange an escort to the counselor's office.
- Monitor social media and insert safe messaging when appropriate.

We wanted to spread the message that we were establishing a "circle of care" around each of our students. This meant working together collaboratively —opening communication channels between students, staff, parents, school and community agencies to keep kids safe. Staff were educated about risk factors and heightened their vigilance; Students were given safety cards with phone numbers for resources and hotlines, and permission to leave class and go to counseling. We established messages of resilience and recovery.

Phase III: Months Later

As weeks turned into months, our school staff and students tried to move through the healing process. We continued ongoing efforts to foster awareness among staff and parents about the contagion risk, educate about risk factors of suicide such as precipitating events (loss, victimization, potential triggers at school or home), and identify warning signs (suicide notes/social media posts, threats, displaying a sense of hopelessness, impulsivity and risk taking behaviors, or giving away prized possessions). The referral process was continually reviewed. However, even with this information, management of the potential for contagion was challenging and anxiety producing. There are no true, completely accurate predictors of youth suicide, and levels of risk can change drastically within hours. It was important to establish ongoing collaboration between counselors, administrators, and me. We regularly conducted crisis team meetings to update our monitoring of students at various levels of risk.

The potential for suicide contagion and the vigilance from staff created another challenge—figuring out how to manage the large number of referrals for immediate suicide assessment. Often we were overwhelmed with students who needed to be assessed and because multiple assessments could be happening at the same time. We often did not have full teams or time to gather best practice information. Our crisis team members' roles and procedures needed to be reviewed and redefined during this phase. We found it was imperative to ensure our suicide risk assessments were collaborative. We learned that it was not only in the child's best interest for safety to gather information from multiple team members, but felt that somehow collaboration also became our liability insurance.

I recall a particularly busy day when every counselor was inundated with students needing support and I was asked to talk with a female student because her English teacher had concerns about some things she disclosed to him. Because there was so much going on I did not have comprehensive information about the concerns. When the bell rang at the end of the school day I asked the student if she wanted to continue to talk or have me follow up the next day. She stated she was ready to go home and we could continue the next day. The teacher later came to the counseling office and disclosed more information that led me to believe she was potentially more at risk than I thought. I became particularly concerned because she was not only in physical proximity to the suicide on campus, but her uncle had recently died by suicide. When I heard the new information, I broke down and became hysterical—probably a warning sign I was overwhelmed and needed to rely on my colleagues to follow through. I could not stop thinking about how I did not have the information I needed or the support of a colleague to determine risk appropriately. A counselor came to my rescue, reassuring me that she would call the student's home, make sure she was home safely, and share the concerns with the parent. This kind of team support has become a staple in our intervention efforts.

We are fortunate in Los Angeles to have a Psychiatric Mobile Response Team (PMRT) to assess kids at imminent risk for suicide and hospitalize them on a 72 hour involuntary hold if necessary. While this can be a valuable resource, it can also be a very time intensive intervention. Our school crisis team involves PMRT only when absolutely necessary to assure the safety of our students so we don't burn out their valuable services.

In total, well over 100 students were referred and assessed for suicide risk. Approximately 25 students were hospitalized on 72 hour holds for their protection. It was noteworthy that the majority of those hospitalized were members of all three designated risk groups. While students who were physically or emotionally proximal to the death appeared at increased risk, observation seemed to indicate that the critical factor in their present suicidal ideation was pre-existing history of mental health issues, suicidal behaviors or traumatic losses.

Phase IV: Contagion/Copycat?

The following summer I remember getting an influx of calls and texts from two different administrators and a counselor that gave me a feeling of alarm. My gut instinct that something bad had happened unfortunately turned out to be right. One of our students had hanged herself in her room after coming home from a party. Then, in the fall of the next school year at the other large public high school in our district, a girl attempted suicide by jumping off a building and survived. She had also reportedly been fighting with her boyfriend. It was then I truly began to worry, not only about the contagion we were experiencing, but also copycats.

Our worst fears were realized. Within weeks of the anniversary of our student's death on campus, a student at our closest neighboring district's high school jumped to his death off a building after school. Social media lit up, reporting that it occurred after an argument with his girlfriend. I happened to have worked part time in this district and was still friendly with former colleagues. I was notified and immediately went to the school to offer resources such as *After a Suicide: A Toolkit for Schools* (2011).

In contrast to our student whose steps were well planned, although the method was the same, this student's actions seemed more impulsive. Also, it was later in the day and the student jumped from a more secluded site. However, the student landed directly in front of the softball field, startling the girls team that was playing a league game. In a terrible twist, they were playing the girls softball team from my school. I was shocked to learn we now had a small subgroup of girls who had actually witnessed both horrific suicides.

While my school crisis team responded over the next few weeks, I focused on this group of girls. Initially they displayed a wide range of crisis reactions. Due to my training in trauma response I was particularly alert to any students reporting intrusive/repetitive thoughts, hyper-vigilance or avoidant behavior such as absences from school.

I was certain the parents of our girls team were going to be very concerned so I started with a parent meeting the National Emergency Assistance Team (NEAT) informally refers to as a Y'All Come—an open invitation is sent to parents for a school meeting. Rich was invited by the superintendent of our neighboring school district to provide guidance and support. He encouraged me to attend the scheduled Y'All Come and I thought it was an excellent opportunity to see how these meetings are conducted.

RECOVERY

Y'All Come

These are critical community forums for parents and I have learned they tend to be well attended. While serving as valuable *caregiver training* during a crisis, when carefully facilitated these forums can also move a community toward the road to recovery. It is a bit of an art to conduct a parent forum as Rich and other NEAT members have told me because there are many ways for the meeting to go "bad". Here are some lessons I learned from attending the parent community forum:

- The forum is advertised as an informational meeting, not a memorial.
- All speakers are public speakers—if they lose their composure, there is potential to expose the entire audience to trauma.
- Representing the district is the superintendent who introduces the principal.
- The principal provides the latest crisis information, what interventions have been implemented on the campus, and how parents can access services.
- Local suicide prevention experts or mental health professionals describe to parents how they can talk with and support their children.
- Parents are informed and empowered to take their *gatekeeper* roles, knowing warning signs and when to refer.
- *Risk factors/warning signs* are reviewed and the referral process detailed for concerned parents and other caregivers.
- Parents are provided handouts on mental health resources, community agencies and services such as American Foundation of Suicide Prevention, National Alliance on Mental Illness, and the National Association of School Psychologists.
- Parents are advised to monitor social media and provided information on *safe messaging* when talking about suicide.
- A lengthy period is left to answer questions written by parents on index cards collected by the school crisis team members and reviewed before

posing to speakers. If parents are allowed to express questions through an "open microphone", there can be many problems that take the meeting off track.

- Keep it brief, 90 minutes to 2 hours max.

LESSONS LEARNED AND REFLECTIONS

Postscript to Crisis: Where Are We Now?

I continue to disseminate messages when I can about the importance of suicide prevention and intervention in my school and the surrounding community. I try to help my district send messages that are accurate and hopeful about suicide prevention. I advocate when I can that suicide is preventable, there are treatments for risk factors, that kids are resilient and recovery is possible, and that everybody plays a role in suicide prevention. We have ongoing discussions, which are particularly valuable for the constant influx of new members at my school, to reestablish roles during suicide threat assessments and review procedures. My school has a strong team that works together to assess and intervene when students are at risk. However, it is noteworthy that, of the counseling/administration staff at the time of the death by suicide, only four of eleven are the same members almost five years later. Whether burnout, secondary stress or natural transiency in school staffing, the turnover of people always presents challenges to maintaining and caring for crisis teams. This reinforces the need for teams to meet every fall and start from the beginning!

Writing this personal narrative has been a profoundly useful journey for me—a validating reflection about my professional growth. I am proud of all the hard work I have put into supporting students and learning about suicide prevention and postvention. I continue to work at the same high school and am in my sixth year. I hold a strong commitment to my school, the staff, students, and families that is strengthened by what we have been through together as a school community. There are, however, times I struggle with the stress. Recently, a staff member burst into a conference room where we were having a meeting and demanded that the administrator help immediately with a situation on the quad needing critical attention that she described as "a mess." I felt transformed through time to that day almost six years ago.

There are some amazing and resilient kids that I've been lucky to intervene with and support through the years. Michael was hospitalized for suicidal ideation three years ago and went through a tough time, but seems to be in a very stable place now. He comes back to visit me on breaks from college and reports he is happy and doing well. We have a great rapport and bond. Jason knew the student who died by suicide on our campus and struggled with guilt and his own academic and personal issues in high school. He also keeps in touch with me and has matured and entered a new phase in his life. Jennifer is a current student who knew she needed help for her suicidal thoughts and reached out to her counselor. We are currently rallying behind her and providing ongoing support. Sometimes I wish I could go back and tell that younger me who was so stressed about finding a job that I would be doing rewarding work with a team of people who would save lives. But I guess learning and growing professionally in areas you never expected is a natural part of the difficult job of a school psychologist.

Special Considerations

Here are just a few more reflections on these events:

- Overall I was happy to be part of a supportive team and couldn't imagine what it would be like if I had to make all the decisions. However, it was frustrating at times when I saw bad decisions or no decisions being made along the way.
- I was very fortunate to have been able to connect with colleagues through NASP and to have subsequently developed a relationship with a mentor who not only guided me through best practices, but also helped to encourage and validate me throughout the process.
- Crises bond responders. Invaluable bonds are formed with colleagues during crisis response. My mentor is still incredibly valuable to me and I have stronger, deeper relationships with colleagues who went through this experience with me at my school and in the district.
- Psychological Triage was critical to help guide individual interventions— once we worked out the Triage Interview, it was easier to sort student need by exposure. Triage is an ongoing process and we were constantly assessing student needs.
- Collaborating with community agencies and emergency services is key, especially emergency mental health personnel. It was incredibly helpful to have them on campus the first few days.
- Tracking identified students, particularly those who were hospitalized, was challenging, but vital—it is important to have a *reentry* meeting and safety planning for each student returning from mental health hospitalization.
- I believe we could have reached out more to staff. Every response should begin with triage of staff as their reactions can impact those of their students. This was a 10th grader who had a total of 12 teachers over his two years at the school. There was also impacted staff from the middle school. I learned the benefit of personally communicating the death of their student in person rather than notifying them through a staff memo or email.
- Utilize your custodial staff to establish policies and procedures to restrict any access to a school roof or other dangerous areas.

REFERENCES

American Foundation for Suicide Prevention and Suicide Prevention Resource Center (AFSP & SPRC). (2011). *After a Suicide: A Toolkit for Schools.* Newton, MA: Education Development Center.

Brock, S., Reeves, M., Nickerson, A., Jimerson, S., Lieberman, R. & Feinberg, T. (2009). *School crisis prevention and intervention: The PREPaRE model.* Bethesda, MD: National Association of School Psychologists.

Lieberman, R., Poland, S., & Cassel, R. (2008). Best practices in suicide intervention. In A. Thomas & J. Grimes (Eds.), *Best practices in school psychology* V (pp.1457–1473). Bethesda, MD: National Association of School Psychologists.

11 Critical and Emerging Needs in a Suicide Cluster

Lessons in Crisis Response, Long-term Planning, and Coordination

Cynthia Dickinson, Richard Lieberman, and Scott Poland

INTRODUCTION

Fairfax County Public Schools, the tenth largest school district in the United States, has more than 186,000 students. The teen suicide rate was lower in the county than in the state or nation, but this did not predict a suicide cluster that occurred within a three year period, alarmingly, within the same high school community. This suicide pattern evolved from 2011 to 2014, raising concerns about the need for effective prevention, access to resources, and response capacity. Over time, it became clear that the local crisis team, and those responding regionally, needed to address long-term, unmet needs within the school community. It became essential to examine each response and postvention plan carefully and thoughtfully, increase available school mental health support, and reengage community partners, including parents and caregivers in providing increased access to mental health support services.

DESCRIPTION OF RESPONSE

Initial Responses

The first teen suicide, January 2011, was later determined to be part of a pattern or cluster. It occurred in winter, during a very difficult period for his family, when the youth was struggling with a change of schools. He had recently left the high school in our county. He was an athlete and Boy Scout, and had experienced what would later be termed a "discipline crisis". After the death, his Boy Scout troop health and wellness leader, a practicing physician, asked regional crisis team members to meet with the scouts to discuss youth suicide, traumatic grief, and where to seek resources. School crisis team members met with affected peers requesting support at his former high school, but that intervention had not reached the Boy Scouts, still reeling from the loss. While support was available for the school, the student's former principal did not request intervention. A crisis team "post-response evaluation session" was convened in the aftermath, and as crisis team manager, I participated in an extended dialogue about the effectiveness of the response, with plans to incorporate best practices in future responses. There was disjointed communication within the school, leading the local crisis team to react, not respond, to student needs. The principal had struggled with the idea of involving his team, as the student was no longer enrolled in his building, and chose to allow support needs to emerge without a comprehensive crisis plan. There were no classroom meetings, no general announcements or talking points for office personnel, administrators, or crisis team members, and there was no coordinated outreach to his peers, though he was still very much part of the student body. Without critical direction, talking with students about the suicide of a former student was quite challenging, and not considered an option. At that time, the outstanding resource, *After a Suicide: Toolkit for Schools* (AFSP & SPRC, 2011) had not yet been released, so those best practices were not implemented. I left the meeting thinking that the response could have gone better, and that the work lacked coordination.

The school community was feeling the pain of the loss, along with the grieving family, who chose to make the disciplinary-connected factor related to the student's death a platform for district-wide reform. Suddenly, any severe disciplinary measure, e.g. referral for expulsion, and/or administrative placement was measured by whether this practice would increase suicide risk. Even the press picked up on the theme of disciplinary crisis leading to suicide, using this student's case as an example.

In August 2012, a second teen from the same high school died by suicide. He was a military-connected student and a student athlete. The family had returned to Fairfax County Schools from another assignment. He and his twin brother were on the same team and summer practices were underway. The district crisis team members met with students and coaches to devise practical ways to intervene, as school was not in session. Crisis team members talked with the team as a group, and in smaller groups. We provided standard messages, and talked about how to seek care for self or others when tragic deaths occur. We discussed risk factors, and we talked with the coaches about appropriate memorial activities. The student's class sponsor and one coach were able to identify students who might need special attention, so coordinated outreach was in place. We publicized and offered school mental health appointments for students who needed individual follow up support. One challenge of the response was that the much-needed follow up planning occurred without the school-based crisis team members being available. Critical follow up with local caregivers—the crisis team at the school, had to wait until the staff members returned to work weeks later. Another challenge was that parent notifications about the availability of resources at the school went unnoticed, as many families were out of town, and not accessing high school email messages. Parents later complained about the lack of timely, accessible information about how to help their children.

Just two months later, October 2012, a student athlete new to the school disappeared. He had driven the family car to school, dropped off his brother, and gone off campus. Homecoming was scheduled that weekend, and he was still not located by the time of the game. The district crisis team was called to attend a vigil after the homecoming game, hoping that the student was lost and unable to reach home. The worry spawned a rumor that he had been kidnapped or murdered. Sadly, his body was discovered by a volunteer search team in a local park the following Monday. His brother, a Boy Scout, had organized the vigil and the search. The collective community felt this traumatic loss and tried its best to cope. Once the sad news was conveyed, the district crisis response team was deployed to assist the school crisis team in a comprehensive response.

Unbelievably, another high school youth died by suicide in April 2013. This young man's parent died the previous year. He was experiencing a decline academically, and was seeking help. Unfortunately, the help came too late. He had been a long-time resident and had a younger sibling at the neighboring middle school. The district team offered comprehensive support, and made certain that the grieving family received appropriate care. Additional support was offered to youth and families who had lost family members to suicide and teachers who had lost students to suicide, as the team realized that hearing about another suicide was traumatic. By this time, teens, parents, and teachers began to question whether something was seriously wrong with the school community. I relayed the idea that the school and its feeder schools were in need of a very large "stress inoculation", as the level of fear, concern, and traumatic exposure increased.

As if the level of trauma for this community was not enough, in February 2014 two more students died by suicide, just 36 hours apart. The crisis teams, particularly the school team, gathered to triage, assess, and implement well-practiced interventions. The shock and trauma of two more deaths within a short period of time was palpable. Team members tried to navigate critical needs for student support, care for caregivers, and comprehensive response. Snowfall and a school closing intervened over the weekend, delaying direct outreach to students, teachers, and parents. All weekend, text messages and emails conveyed an increased level of concern about affected students. There was no easy way to set up a call center, so crisis support information, including messages about how to talk to students about traumatic events, was posted on the school website and sent out via social media.

I was struck by the feeling of powerlessness and helplessness that we, the mental health team, were experiencing. Our own feelings of shock and disbelief began to erode our collective sense of self-efficacy. I knew that we had quickly reached the point of needing outside help.

A community meeting, convened by the superintendent of schools, and attended by leaders from the school district, county government, police and health departments, parents, teachers, and youth was held in early March 2014. A small information fair was scheduled afterward in the school cafeteria and was mobbed by parents, educators, and students. They wanted reassurance, help, and hope. The superintendent made a public commitment for a mental wellness summit, so parents and community members could talk about unmet mental health needs, available resources, and better ways to collectively address this ongoing crisis. She also allocated funds to provide more mental health services and administrative support for the remainder of the year. It was obvious that our district needed to retool and work on more deliberate communications and support to the school community. We had to think comprehensively. We needed to call on trustworthy "outside experts" to ensure our work was on track. The cost was simply too great.

Sources of Support Throughout the Responses

Often, in the midst of serial crisis responses, it is difficult to gain perspective about the incidents and their impact. Even before this alarming pattern of youth suicides was evident, the school crisis team was asking, "What's the connection among these youth? They reside in the same community, are male, and some are from military families. Some are athletes, but not in the same sport. Some are scouts, while others are not. The only connection known was that they were struggling somehow, and that they attended the same high school." An even more concerning note, none of the methods of self-inflicted harm were similar.

It took conference calls with National Association of School Psychologists (NASP) crisis experts to confirm the presence or absence of a "suicide cluster" or another phenomenon. A call to Dr. Stephen Brock, PREPaRE (2009, 2016) co-author and School Safety and Crisis Response Committee colleague, helped our leadership team refine its perspective on "what constitutes a troubling pattern, such as a suicide cluster", the related emotional and physical proximity factors, and determine a plan forward. A call to Frank Zenere, National Emergency Assistance Team (NEAT) emeritus member, PREPaRE trainer, and crisis manager, elicited information about safe media reporting, and the importance of providing information about available help and resources. We also discussed the possibility of poor media coverage adding trauma to responders. I assured him that responders were being adversely affected. The coverage continued with front page news of the suicide cluster in the Washington Post. The reporters failed to follow responsible reporting guidelines, telling the painful family stories without directing readers toward help.

I placed calls to the Substance Abuse and Mental Health Services Administration (SAMHSA) to confer about available suicide postvention resources, and to whether the Centers for Disease Control and Prevention (CDC) could come in to address this array of crisis incidents. We discussed the SAMHSA High School Suicide Toolkit, and hotline resources. I also located a Canadian mental health researcher who had conducted a recent study on suicide clusters, and asked her for a copy of the study and recommendations for communicating about available resources within school settings.

Taking Action in the Midst of the Crisis

On March 5, 2014, the superintendent's recent call to action mobilized a mental health summit, so that parents, youth, and community members could discuss civic challenges, wellness programs, and ways to support each other and seek help. Community service providers partnered with school system members, and presented talks on mental wellness, coping strategies, resources, and a keynote focused on stress and resilience. An exhibit hall provided the opportunity to learn more about community and school district resources. This event afforded parents, students, school and private mental health professionals, and educators the opportunity to discuss the physical, psychological, and academic pressures affecting youth, and to offer a forum to discuss available behavioral, health and educational resources.

While the mental health summit met a set of needs, it was not located at the affected school, and many of those families did not attend the event. Conferring about so much trauma and loss was insufficient, so in the fall, 2014, I applied for and was awarded a Project School Emergency Response to Violence (SERV) grant through the US Department of Education.

Project SERV: Expertise on the Ground

The grant funded additional clinical support at the school, a student support team coordinator for discussion of at-risk students, and consultation and training by founding members of NEAT, Richard Lieberman, formerly the Suicide Prevention Coordinator for Los Angeles Unified School District and Dr. Scott Poland, Co-director of the Suicide and Violence Prevention Office at Nova Southeastern University.

In three tightly scheduled days, Rich and Scott met with administrators, parents, teachers, students, and crisis team members, and provided best practice education on youth suicide prevention and intervention. Among the many goals were to get everyone on the same page, review our policies and procedures, and provide a sense of hope that our community was not alone and we were universally committed to preventing the next suicide.

A parent community forum, incredibly well attended, was designed to keep parents informed of the district's commitment to prevention activities and empower parents to their critical gatekeeper role in limiting contagion.

A lengthy time was allowed for all questions to be answered by the speakers. Many commented that the information was helpful for parenting a teen in today's world. They admitted initially dreading the meeting, thinking it would be too depressing, but reported afterward feeling hopeful and empowered to stay involved in their child's life with a better understanding of mental health.

All faculty from the affected high school and the feeder middle school were given time off to attend a separate afternoon presentation that focused on the staff role in suicide prevention. They were reassured they could talk to their students about suicide without any fear of putting ideas in their heads. The risk factors and warning signs of suicide were reviewed and the referral process was clearly defined. Here too, all staff questions were addressed.

The suicide prevention presentation to administrators included lessons learned from lawsuits brought against schools after a suicide. Scott and Rich were personally involved in many of these lawsuits and stressed the important lessons for prevention, including essential ways to protect youth so that mental health emergencies receive appropriate, documented follow up. One school administrator told Scott that it did not sit well with her that he had been on the side of the plaintiff in a lawsuit against a school district. Scott responded that his responsibility was to describe the need for suicide prevention protocols for high risk students! It was notable that no elementary administrators attended. Unfortunately, younger and younger elementary students are dying by suicide. It would have been ideal if the superintendent had attended one of Scott and Rich's presentations, but the deputy superintendent attended the administrator session and continues to support ongoing district wide crisis response and prevention efforts.

Most moving for me was the presentation to students, who had a chance to debrief, reflect on their losses, consider academic and social media stressors, and learn ways to help their friends and themselves. Their gatekeeper role was reduced to two simple rules. Never hold a friend's suicide intent a secret, and tell a trusted adult. The students shared a lot about academic stresses they were coping with and admitted they placed much of the stress on themselves. They frequently discussed their class rank with peers and lost sleep over early classes (and time on social media!). Research links sleeplessness as a distal risk factor for youth suicide. Scott and Rich then shared personal stories and emphasized that there are many paths to success and things rarely work out perfectly right out of high school. Scott shared that he was actually kicked out of college initially for poor scholarship and that he is an example that it can all work out.

Before leaving, Scott and Rich facilitated two essential meetings. The first was designed to bring all local and state suicide prevention stakeholders together to share resources, build alliances and support our district. The meeting was well attended, but the medical community was a glaring absentee. Scott and Rich shared from their experience that it is difficult to engage the medical community even after a suicide cluster. Recommendations have emphasized the need for physicians to be more engaged in suicide prevention. Many individuals who died by suicide saw their family physicians before their death. It has been suggested that every teenager seeing a family physician for any reason be screened for depression and suicidal thoughts (Maurer, 2012). This is absolutely essential after a cluster, and this message was reinforced in our commmunity.

The second meeting was a session with our school crisis team. Ostensibly it was to evaluate our response, predict and prepare for the challenges of the next few months, but it seemed to emphasize the impact on the responder and how self-care was a fundamental necessity for all of us to see this job through. Additionally, Scott and Rich offered four interactive webinars for parents, school staff and mental health personnel, and media. These webcasts were archived on our district website and instantly available resources. Topics included "talking to youth about trauma and loss after a tragedy", "non-suicidal injury", "media coverage and safe messaging for prevention", and as an update for our school site teams, "suicide postvention for crisis teams".

RECOVERY AND LESSONS LEARNED

Special Considerations

After the superintendent's community meeting, a small group of parents formed a collective "Community of Solutions" group, which began to address unmet needs within the community. The high school parent and physician, who had requested Boy Scout support in 2011, became a public healthcare policy advocate, seeking ways that primary care physicians could screen youth for mental health risk during school physicals—annually for school athletes, a group of particular concern. He worked on concussion and sports physical protocols, and invited school personnel to "grand rounds" at the local hospital to share data from the annual Youth Risk Behaviors Survey. The Community of Solutions group follows research about youth risk factors, including substance abuse, concussions, and mental health risks, and invites speakers to discuss mindfulness, academic stress, and how to develop healthy coping strategies. Partnerships with agencies, including mental health and substance abuse services have resulted in offering Youth Mental Health First Aid training and free Kognito (gatekeeper) training to the community.

The American Foundation for Suicide Prevention (AFSP) has partnered with the district in supporting mental health summits, survivor groups and support efforts. When families are affected by a suicide, our district now knows the resources to contact.

Suicide clusters are a pattern that the Centers for Disease Control and Prevention wants to study. Our local health department invited the CDC to conduct an Epi-Aid study to look at risk factors, support services, and improvements needed to stem the tide. Its study, published in 2015, found among significant findings that local reporters had not complied with media guidelines for reporting on suicide, and that parents, especially in the affected community, needed help identifying their youth's mental health concerns, and navigating mental health resources.

We have learned from the collective crisis experience within our community, that it is vital to look at what others are doing. Our district is following the work in Palo Alto, California, where local districts have experienced suicide clusters and are convening school mental health, and other community agencies to address the crisis. Palo Alto also had a CDC visit, and an Epi-Aid study within the past year. In spring 2017, several members of our team, with representatives from Palo Alto, CA and Colorado Springs, CO are participating in a pre-conference workshop for the American Association of Suicidology moderated by Scott and Rich. Their goal is to spark renewal into researching a best practice protocol for responding to suicide clusters in which the schools are an essential part of the Unified Command.

Before this series of incidents, school district employees were not part of the county suicide review team. We now participate in this review process, often referred to as a "psychological autopsy", discussing what was known about students who died by suicide.

Evaluation showed that one critical overlooked option in comprehensive planning was the ongoing need for professional and personal self-care for crisis team members. PREPaRE Workshop 2 (Brock, 2011) training affirms that self-care is as vital as having the requisite skills for response. While we did not minimize the need for breaks and support throughout the responses, we often relied on support from the community. Examples include meals and snacks provided for the school mental health providers during responses, and the opportunity for regular processing of the event and planning for the next day. A grant funded support to help alleviate the workload. Dialogue was insufficient about questions like "Where does this crisis place you?" "What are you doing to ensure that you are taking good care of yourself?" "Do you need a break from this?" While we facilitated that type of dialogue with others such as teachers, we did not have that routine for the crisis team, many of whom were just relieved when the work day ended. We have forged a new partnership with our Employee Assistance Program to bring in on-site counselors to address crisis responder and adult staff needs.

As the grant period ended, we engaged in a "lessons learned" dialogue, at the end of the school year. School crisis team members shared what they learned, and also their hopes and dreams for the summer, and plans for the next school year. Some chose to leave the school for a new assignment. Others were looking forward to sharing time with their families, and having time off from a series of difficult years at the school. They emerged stronger, having addressed critical needs within their school setting, and focusing forward on prevention and intervention efforts.

REFLECTIONS AND MORE LESSONS

As I continue to be reminded, this unique suicide cluster occurred during a finite period of time. However, in my large school district, youth suicides continue to occur. We incorporate best practices from the *After a Suicide Toolkit* and the PREPaRE Workshop 2 (2011) training, now standard for all crisis responders, as we address the common experience of youth suicides.

We must not forget that each loss affects a community, even the larger community, and the ripple effects can be widely felt. We continue to implement wellness screenings and partner with our mental health and survivor networks. We continue to develop resources for community groups, including youth organizations, so their leaders can have discussions with youth about ways to seek help, and not see suicide as an option. I continue to believe that our outreach and education efforts will afford our youth a more hopeful perspective, and that community members, including parents, faith leaders, and teachers, will surround our youth with support and care. With the support, consultation, and training by a team of professional colleagues, we have emerged with strength and resilience.

REFERENCES

American Foundation for Suicide Prevention and Suicide Prevention Resource Center (AFSP & SPRC). (2011). *After a Suicide: A Toolkit for Schools*. Newton, MA: Education Development Center, Inc.

Brock, S.E. (2011). *PREPaRE Workshop #2: Crisis intervention and recovery: The roles of school-based mental health professionals*. (2nd ed.). Bethesda, MD: National Association of School Psychologists.

Brock, S.E., Nickerson, A.B., Reeves, M.A., Jimerson, S.R., Lieberman, R.A., & Feinberg, T.A. (2009). *School crisis prevention and intervention: The PREPaRE model*. Bethesda, MD: National Association of School Psychologists.

Brock, S.E., Nickerson, A.B., Reeves, M.A.L., Conolly, C.N., Jimerson, S.R., Pesce, R.C., & Lazzaro, B.R. (2016). *School crisis prevention and intervention: The PREPaRE model* (2nd ed.). Bethesda, MD: National Association of School Psychologists.

Maurer, D.M. (2012). Screening for depression. *American Family Physician*, 85(2), 139–144.

Section 5

School Crisis Response to Natural Disasters

When I was a boy and I would see scary things in the news, my mother would say to me, 'Look for the helpers. You will always find people who are helping.'

Mister Fred Rogers

INTRODUCTION

Definition

Natural disasters are catastrophic acts of nature that can disrupt communities, traumatize people, create scarcity of basic needs, and result in injuries and fatalities. The suddenness and severity of natural disasters determines the degree of impact. They often occur with minimal warning, so prevention, preparedness, and intervention are needed to mitigate emotional distress, property damage, injuries and deaths (Heath, 2014). FEMA.gov and Ready.gov have websites with information on response to specific natural disasters, including hurricanes, tornadoes, floods, wildfires, earthquakes, and tsunamis.

Vulnerability of Children and Adolescents

Children and adolescents are especially vulnerable to natural disasters as the devastation to communities can be frightening, long lasting, and damaging to a child's sense of security. When homes are destroyed, they must cope with living in a shelter or being relocated. Children's ability to cope is influenced by how they see parents, teachers, and others coping. Emotional trauma can be reduced by involving them in developmentally appropriate pre-disaster planning, and by teaching them regulation of feelings, coping strategies, and resilience.

Role of Schools

Returning children to schools and familiar routines can be a vital asset in helping communities begin recovery. Schools can plan in advance to meet students' mental health needs, provide information and resources for staff and families, and help to stabilize and normalize a community in the event of a natural disaster (Heath, 2014; Lazarus et al., 2003). A caring school staff can help reassure children as they return to normalcy. During distressing times, students can find comfort and learn from their teachers and school staff.

PREPAREDNESS FOR NATURAL DISASTERS

Strategies to Prepare Schools for Natural Disasters

- In response to natural disasters, maintaining school routines can have a significant stabilizing effect on distressed communities.
- Schools can provide disaster education and preparation for students and families.
- *Practical Information on Crisis Planning: A Guide for Schools and Communities* is a useful resource: rems.ed.gov/docs/PracticalInformationCrisis Planning.pdf.
- PREPaRE training (Brock et al., 2009, 2016) is outstanding for school crisis preparedness and intervention.
- School disaster response plans should address how to communicate with families, storage of food, water, medical, and other supplies, and plans for "shelter-in-place", reunification with families, and evacuation if needed.
- Schools can plan for public media, social media, and Internet to give parents guidance for post-disaster recovery and helping children cope.
- Schools should plan to meet the physical and emotional needs of students with special needs, especially exiting and relocating those with limited mobility.
- School emergency drills should include community first responders and agencies.

Providing Resources for Students, Families, and Community

- Plan for additional school-based and community mental health providers, and also staffing to provide before and after school care, and possibly weekend care.
- Schools should try to keep students, staff, families, and the media informed by any means available—text messages or email may work when networks such as cell phones and land lines are congested.
- Collaboration between school crisis response teams and a variety of community, state, and federal agencies can bring resources into schools, helping connect families with housing, financial, insurance, medical, or emotional assistance.

- After severe disasters, teachers can provide developmentally and culturally appropriate information such as community recovery efforts and weather reports to prepare, reassure, and alleviate fears.

SUPPORTING EMOTIONAL NEEDS

Support Emotional Needs of Students

- Return to normal school routine as soon as possible, but administrators, teachers, and crisis responders may first need to address student concerns, carefully listen to their experiences, and acknowledge feelings.
- Trained teachers and administrators can provide developmentally and culturally competent school-wide interventions and activities to support psychosocial needs.
- After severe disasters, provide students with ongoing information about when and where to access counseling, and provide appropriately structured opportunities for classroom, small group, or individual discussion as needed.
- Magic circle sharing for children, and guided group discussion for teenagers at the beginning of the school day can prepare students for school work, dispel rumors, and identify those who are upset and need referral.
- Caregivers should remain calm and reassuring, acknowledge destruction and distress, but emphasize efforts to care for people and rebuild the community.
- Monitor children's use of television, Internet, and social media, minimizing exposure to frightening images, and educating them about media coverage.
- Plan for possible student fatigue due to stress reactions by reducing instruction, providing snack time for extra nourishment, and postponing tests if needed.
- Return to academics as soon as possible, but consider integrating relevant disaster information into subjects such as science, math, history, and language arts.

Support Teachers and School Staff

- Plan opportunities for school staff to receive support and share reactions so they are emotionally stabilized and better able to support their students.
- Teachers can honestly display sad feelings, but must maintain emotional composure and control of the situation to reassure students.
- Teachers, with the support of crisis responders and school-based mental health providers as needed, can structure a classroom responsive to student reactions.

- Plan for floating substitutes in case teachers need breaks, time for family disaster planning, or crisis counseling.

Involve Students in Meaningful Activities

- Restore a sense of control by involving students in activities like charity work for survivors, writing cards for grieving families, writing poems and stories to share, or drawing pictures or murals.
- Consider forming an advisory committee of students with adult guidance, to help identify needs and plan resources to support coping and recovery.
- Recognize constructive coping and teach resiliency, problem solving, and strategies to manage disaster-related stress.
- Listen to children's concerns, stories, and questions, normalizing their feelings, addressing their concerns, and involving them in developing safety plans.
- Encourage, but do not force children to talk about their disaster experiences, or express themselves through art, play, writing, or other activities.

Continue Monitoring for Students Needing Emotional Support

- Identify students needing support, and channel toward crisis counseling and other resources if needed.
- Children may express a range of emotions, but be prepared to address typical fears after a natural disaster—that the event will happen again, someone close will be injured or killed, or they will be separated from family and home.
- Help children and adults understand that sadness, grief, fear, and anger are normal reactions to a disaster.
- After severe disasters, be vigilant for long-term reactions such as on anniversary dates, and refer those having severe reactions for treatment.
- While evaluating student needs, document contacts, inform parents of those at risk, and make treatment referrals when needed.
- Carefully evaluate needs of students whose homes were damaged or destroyed.
- Parents, educators, and caregivers can use social support systems, including family, friends, school, faith-based organizations, and community agencies to help children and adolescents cope with a disaster.

Support Displaced and Relocated Students

- Receiving schools can provide displaced students with needed stability, especially when displacement is expected to be long-term.
- Encourage children's friendships and peer support, especially maintaining relationships disrupted by relocation.

- Determine addresses and phone numbers of all relocated students, encouraging interested classmates to write or phone.
- When relocated or in temporary housing, bring children's valued personal items and be especially sensitive to their needs.

Closure and Long-Term Support

Planning follow-up for severe stress reactions:

- While mental health resources tend to be plentiful immediately after a traumatic event, they are generally less available weeks, months, and years after the event, when those most affected may still need support.
- Although recovery is the norm over time, children and teenagers continuing to suffer from stress reactions may need referral for therapeutic treatment.
- After severe disasters, monitor for long-term and delayed stress reactions, and risk for suicide indicating immediate need for treatment referral.

Planning for funerals and memorial services:

- In disasters involving deaths, plan for funeral and memorial services, including respect for the family's wishes, choice to attend or not, having parents accompany their children, and having school staff monitor and support students.

Caring for caregivers:

- Caregivers must care for themselves, seeking support from colleagues, friends, family, faith, or when needed, from mental health providers in order to continue caring for children.

Adapted from Brock et al., 2009; Heath, 2014; Lazarus et al., 2003; Nastasi et al., 2011; Roth, 2015; Zenere, 2007

REFERENCES

Brock, S.E., Nickerson, A.B., Reeves, M.A., Jimerson, S.R., Lieberman, R.A., & Feinberg, T.A. (2009). *School crisis prevention and intervention: The PREPaRE model.* Bethesda, MD: National Association of School Psychologists.

Brock, S.E., Nickerson, A.B., Louvar Reeves, M.A., Conolly, C.A., Jimerson, S.R., Persce, R.C., & Lazzaro, B.R. (2016). *School crisis prevention and intervention: The PREPaRE model* (2nd ed.). Bethesda, MD: National Association of School Psychologists.

Heath, M.A. (2014). Best practices in crisis intervention following a natural disaster. In P.L. Harrison & A. Thomas (Eds.) *Best practices in school psychology: Systems-level services* (pp. 289–302). Bethesda, MD: National Association of School Psychologists.

Lazarus, P.J., Jimerson, S.R., & Brock, S.E. (2003). *Helping children after a natural disaster: Information for parents and teachers*. Bethesda, MD: National Association of School Psychologists. Available online www.nasponline.org.

Nastasi, B.K., Jayasena, A., Summerville, M., & Borja, A. (2011). Facilitating long-term recovery from natural disasters: Psychosocial programming for tsunami-affected schools of Sri Lanka. *School Psychology International, 32,* 512–532.

Roth, J.C. (2015). *School crisis response: Reflections of a team leader*. Wilmington, DE: Hickory Run Press.

Zenere, F. (2007). Lessons learned from recent hurricanes: Efforts related to schools and students. *NASP Communique, 35* (5).

12 Children of Katrina

A View from Texas and Mississippi

Gabriel I. Lomas and David J. Denino

INTRODUCTION

August 29, 2005—Disaster Strikes

Lomas: It was a typical day on the Gulf Coast for me. The air was hot and humid, with ominous clouds in the sky. It rained heavily for periods of time, but the weather was nothing serious for Houstonians, who went about their usual business. While adults went to work and children went to school, we were all concerned about the storm that was beginning to pound our neighbors in New Orleans. It seemed certain that the storm would only marginally impact the Houston area, but the greatest impact came in the weeks and months after the storm. This was a significant transition time for me. Since 1996, I had worked in Texas public schools in various roles, primarily as a teacher then as a special education counselor. The new school year was just underway, but I was not with school-aged students. Instead, I reported to my new job as a professor at the University of Houston—Clear Lake. I knew my former colleagues, experts in handling school crises, would respond to challenges that the storm was to bring.

Denino: From my home in Connecticut, I watched Hurricane Katrina unfold with a great deal of concern for fellow Americans and friends along the gulf coast. I had been traveling to New Orleans since 1978 as a favorite destination for music, culture, and food. The call went out for all types of first responders to help with a catastrophic event. At the forefront were first responders from the US Coast Guard, National Guard, police, fire, and other rescue personnel to help with securing the safety of residents. Second responders are also critical, as we provide resources for the basic needs—food, water, and shelter. My connection was as a Red Cross volunteer trained in Disaster Mental Health with a willingness to help both at the local and national level.

DESCRIPTIONS OF RESPONSE

First Call

Lomas: Perhaps my employment transition moved me from direct services to students to the periphery, as a consultant. Though highly trained, I was not registered as a responder with any formal organization. I was hearing first hand stories from colleagues at the university, from graduate students enrolled in my program, and from former colleagues in area school systems. Media reports told of the chaos in New Orleans. Thousands of people were evacuating, and many would end up in the Houston area. Some evacuees came intentionally, finding shelter with friends and family. Others came to Houston unaware of their destination when they boarded busses. For me, the "first call" was an internal pull to help. I knew I should not attempt to insert myself into a crisis unless my presence was requested. Initially, I listened, watched, offered assistance, and provided consultation and supervision to my graduate students involved, as well as peers from the community who reached out to me.

Denino: The Red Cross and the American Counseling Association sent out calls for mental health responders. All responders who were part of the team were trained and vetted by the Red Cross prior to deployment. Proper training and vetting of responders is a key aspect of a well-run crisis response. I was up at 4:00 a.m. and flew to Atlanta, connecting to Montgomery, Alabama. I had no idea where I would sleep that night. I anticipated the accommodations would be challenging, but was unsure what was ahead. By 11:30 a.m., I was in Montgomery and engaged in my check-in process. This included general orientation, assignments, obtaining identification, and screening at the staff health center. While there were hundreds of volunteers arriving with me and behind me, the Red Cross processed each of us with surprising efficiently. All 25 of the mental health volunteers were initially assigned to work in the shelter in Montgomery.

Subsequent Days

Lomas: Evacuees were pouring into the Astrodome, near downtown Houston. In some ways, the Astrodome was ideal as it is large, has many bathrooms, and was no longer used for sporting events. However, the facility was not designed to house people. There was little to no provisions for privacy, and there were no playgrounds for children. Evacuees were exhausted and distressed. Although they were safe from the storm, many were unsure how they would meet their basic needs. Thus, the environment was threatening to some, and there was a need for increased security. Volunteers worked tirelessly to set up separate spaces in the arena. It was important to give children opportunities to play, and to give information to adults who were eager to get their lives back in order. State agencies were heavily involved with evacuees at the Astrodome.

Many children were distressed and some were separated from caregivers, so Child Protective Services was also heavily involved. The Red Cross also had volunteers dedicated to the task of reuniting children with their families.

Schools in central Houston were inundated by large numbers of students displaced by the storm. Counselors and other helpers were busy enrolling them. It is critical for people in crisis to return to homeostasis, and school attendance provides essential structure for young people. Schools in suburban areas were also feeling the impact, though not as significantly. Initially, schools were doing well and most were happy to provide temporary help for distressed children and families.

Denino: In intensely hot and humid conditions, hundreds of people were standing in line, all waiting to be processed by the Red Cross for disaster relief. At this point, children were not in schools. Mental health workers were asked to give one Mickey Mouse stuffed animal to each child. The stuffed animals, each with a Red cross emblem, served as an icebreaker for conversations and as an object of comfort and security. The stuffed animals were a bridge to parents, who then gave us permission to speak with their children. Often, traumatic stories unfolded as people began telling their stories. People of all ages, from children to seniors, were evacuated to Montgomery, each with their own compelling story. As we met with people, we only kept a tabulation of numbers, so the Red Cross could keep a count of how many people we saw in a day. In disaster mental health, we do not take notes in the traditional manner for counseling. We were trained to engage in Psychological First Aid (PFA). I did this for two days, before moving to a new shelter in Brookhaven, Mississippi.

About a Week after Impact

Lomas: The area within and around the Astrodome was still chaotic. Evacuees were frustrated, as many were not yet moved to temporary housing. The Astrodome had become increasingly uncomfortable and dangerous. A large number of children were enrolled in local schools at this point. There were reports that students from gangs were interacting with rival gangs. Some youth from New Orleans gangs were confronting youth from Houston gangs. The consequences for schools were devastating. There were fights that were too large for building administrators, and necessitated police intervention. This unfortunate situation emphasizes the need to try to prepare and involve students and staff from receiving schools to accept and support displaced students.

In one suburban school system, the school-based mental health professionals used the National Organization for Victim Assistance (NOVA) model to process the experience with students and families. NOVA has a group model used to allow individuals to process experiences and mitigate the effects of trauma. Bringing people together in a safe atmosphere usually helps, providing a forum to vent feelings, learn coping skills, and find they are not alone in their struggles. Many school systems in Texas used the NOVA model at that time, or had a similar model. The model gives facilitators an opportunity to screen for poor coping, and refer those needing more intensive help. The timing requirements for groups runs between 60 and 90 minutes, which is generally consistent with a high school bell schedule. Most students appreciate the opportunity to voluntarily share their crisis experiences in a safe and structured setting.

Denino: There were too many people rushing the crisis workers, so the National Guard had to step in and restore order. This was a scary time for the volunteers. We were told we would be moved to a new shelter the next day. For today, I was told to direct traffic on the road near the shelter. At night, we showered with a garden hose rigged from pipes above, wooden pallets below, and a tarp surrounding the structure for privacy. It was so refreshing to cool off. After a long, hot day of volunteering, the makeshift accommodations felt like a spa!

The next morning, we were sent to a new shelter in McComb, Mississippi. We were expecting 10,000 cars with families to come through and get tickets with a return date for an appointment for disaster relief processing. We gave out all 8,000 tickets by 4:00 p.m., leaving many people feeling excluded and upset. Eventually, we got to everyone. Our site was the county fairgrounds. We were tasked with changing large, empty buildings into comfortable rooms. We went to a hardware store and purchased lumber, carpets, paint, and other supplies to help make evacuees feel welcome. The next morning, we woke to 200 people in line for their appointments. We processed all 400 families who had appointments that day. Our work was both task and process oriented—gathering information, assessing needs, and using Psychological First Aid skills when we saw the opportunity and the need. We sat with people and waited, listened, offered care, and extended support. We rotated in and out of assistance stations all day. In each station I heard about multiple losses. People lost their homes, their jobs, and their loved ones. Notwithstanding the trauma and distress, this was a good, very productive, satisfying day. It seemed like everyone I met said, "Thanks for being here—God Bless you and your family". I was amazed to see so many people who lost everything, and yet were kind enough to be thankful. After several days of waiting, children in our shelters were restless. They lost their schools, their playgroups, their extracurricular activities, and the structure that gave their life security and meaning. When it was time for me to go, I took only the clothes I was wearing. I went to a local church, washed everything I brought with me, and had the church give my clothes to people in need.

LESSONS LEARNED AND REFLECTIONS

Lomas: By the time Katrina hit the Gulf Coast, I had already worked for a number of years as a special education counselor and in a part-time private practice. At that time, I didn't realize the gift of NOVA training. Crisis preparation can't be done on the spot. It must be done in advance, and responders must meet and exercise regularly. Today, there are several helpful models to prepare second responders to handle crises. The NOVA model is designed for broad disasters, but the National Association for School Psychologists (NASP) has the PREPaRE curriculum, which is designed for school crisis response. Currently, I work in western Connecticut, only a few miles from Sandy Hook Elementary, the scene of a horrific school shooting to which I responded. I have seen crises, large and small and I know that the best preparation is a well-trained crisis team. Regardless of the selected model, training and exercising in crisis response is essential for you, your school, and your community.

Denino: After being discharged, I was happy to be home. I had so much to process, having woefully underestimated the personal impact the trip would have on me. Crisis response work is exhausting, both physically and emotionally. My experience validated that much of what we observe and do as disaster mental health volunteers is consistent with training like Psychological First Aid (PFA), NOVA, and PREPaRE. I used PFA because I'm trained in the Red Cross Disaster Mental Health model. Here are some basic concepts, (based on FEMA and Red Cross), to keep in mind when helping children cope with disasters.

Understand some Reactions Young Children Experience:

- A return to earlier behavior, like thumb sucking or bed wetting;
- clinging to parents or adults;
- nightmares or difficulty sleeping;
- difficulty with emotional regulation—crying, screaming, or hypervigilance;
- withdrawal, flat affect, or loss of interest in usual activities;
- refusal to attend school.

Adults can Reassure Children's Sense of Security and Model Resilience:

- Keep routines and schedules as normal as possible.
- As an adult, maintain control of the situation to the extent possible.
- Encourage children to talk, listen to what they say, address stated needs, and honestly reassure as often as needed.
- Include children in developmentally appropriate, constructive recovery activities.
- Help children cope by normalizing typical reactions and teaching stress management skills.
- Be prepared to refer for more intensive help if severe reactions continue.

Resources:

NASP: www.nasponline.org
NOVA: www.trynova.org
FEMA: www.fema.gov
Red Cross: www.redcross.org

13 Anticipating Needs after Katrina

Welcoming Children and Families

Jeffrey C. Roth and Crisis Team Members

This memorandum was developed for the Brandywine School District, Delaware by Jeff Roth, Coordinator/School Psychologist, Doug DiRaddo, School Psychologist, Nancy Carney, School Counselor, Kittie Rehrig, School Social Worker, and Elliot Davis, School Psychologist to prepare for inclusion of students and families after the Hurricane Katrina disaster. The information may be applied to other natural disasters.

School District Memorandum in the Aftermath
of Hurricane Katrina

DISASTER RESPONSE: ANTICIPATING THE NEEDS OF STUDENTS, FAMILIES, AND STAFF

Our community is beginning to receive survivors of Hurricane Katrina. It is likely that many of our schools will be providing care and education to students and families from the disaster area. The catastrophic conditions many of these children and families have experienced are truly horrific. Many of them will arrive as "homeless" people, traumatized, without material possessions in a culture different from what they have known. We have an opportunity to educate ourselves and take steps that will help to meet their needs and invite them into our school community.

- Most students will be registered in our schools as "homeless". They may be entering new schools without documentation such as birth certificates or immunization information. In these cases, consistent with the McKinney Act, school policies and practices must assist homeless students to enter schools without delay.
- There is a likelihood that some of these students and families may develop traumatic stress reactions. They will most certainly be coping with grief and loss. Given their emotional distress, these students may have extreme difficulty focusing on academics. They will need a patient, supportive,

non-pressured approach. There must be a balance between respecting the primacy of emotional needs while encouraging a return to a normal academic routine as soon as it can be tolerated. This return to normal routine may vary from student to student.

- **Feeling accepted and valued by fellow students and staff will help students cope and foster the healing process**. The student and family's healing may be aided by connecting them to appropriate support networks in the school and community. Schools offer psychologists, counselors, nurses and social workers trained to support mental health. Social workers may connect families to agencies in the community. The [Name] School District has a Crisis Response Team that is available for consultation and intervention at the request of the school.

- Sometimes the physical, emotional and cognitive symptoms of a stress reaction appear immediately after the event, sometimes not for days or weeks after the troubling incident. Often, people think there is something wrong with them for having unusual and unfamiliar feelings related to the stress reaction. Really, they are having normal reactions people experience after a traumatic event. While people may feel very different from their normal selves in the aftermath of the traumatic event, their "built-in" capacity to recover from upsetting events helps them heal, especially if they are cared for and take care of themselves. Generally, these reactions are time-limited and will diminish.

- **Consider emotional needs of students and staff not directly involved in the disaster**. Media coverage of the horrific events brought on by the hurricane and its aftermath has touched many adults and children. Students in particular may need reassurance from adults that they are safe and "cared for". They may be reminded that many people are helping those in need and that they are being helped in many ways. Children and adolescents may be given some power and hope in this situation by being given an opportunity to express themselves through writing and art, or to participate in acts such as fundraising with adult supervision, to help those directly affected by the disaster.

- It is recognized that school staff members may also be dealing with their own emotional reactions to the disaster. Remember that in order to take care of students, school staff members must take care of themselves. Supporting colleagues and exercising stress management strategies is recommended.

- Advice to children/adolescents about what helps during and after a crisis:

    - **Talk with others about what happened and how you feel about it**. Talk with parents, teachers, friends or family members. When you share your feelings with others, you'll probably realize you're not the only one who's feeling upset.

– **It's okay to think about the upsetting events.** Don't fight recurring thoughts or memories—they are a normal part of recovery and will diminish over time.

– **Use your teachers and counselors at school for help too.** Sometimes it's hard to concentrate in class or talk with other kids about what's happening, so let people know so they can help.

– **It's still okay to have fun.** Sometimes after a bad thing happens, it may seem like we shouldn't be our normal selves, and laugh or have fun. However, it is important now to get out energy and do things you enjoy. Remember, you're not doing anything "wrong" by doing things you enjoy and having fun with friends.

– **Take care of your health.** Make sure you are eating and sleeping okay. If you are having any problems, tell your parents or someone you can trust. Try exercising and getting outdoors. Both are great ways to help relieve stress. Avoid alcohol and drugs.

– **Give yourself time to get better.** Don't rush recovery or have unrealistic expectations you should "get over it" or "shouldn't" be feeling that way. They may be difficult because it takes a while to feel better. If you're too upset to go to school or still feeling really bad, let your parents or someone you trust know. If it seems like you need some extra help, your parents or teachers can get you help outside of school. Remember that everyone handles stress differently. If you're doing what you can with the help of your parents or loved ones, friends and school staff, you will eventually get better.

• Here are some excellent websites with more information on a variety of disaster related topics:

– National Association of School Psychologists (NASP) www.nasp online.org

   ◦ Click on "Katrina Resources"
   ◦ Select topics such as:

      – Responding to Hurricane Katrina: Helping Children Cope
      – Responding to Hurricane Katrina: Helping Students Relocate and Supporting Their Mental Health Needs
      – Responding to Hurricane Katrina: Information for Schools

– APA Help Center **http://www.apahelpcenter.org/**

   ◦ Managing Traumatic Stress After Hurricanes

School/District Letterhead   September 9, 2005

Dear Parents/Guardians,

Over the past several weeks our children have been exposed to images, news and information about the disaster along the Gulf Coast. When children are exposed, particularly to images of events such as **Hurricane Katrina**, they can be particularly vulnerable. They may have difficulty processing the extent of the loss of life, destruction of property, loss of communities and the breakdown of the civil order system. Parents, along with other caregivers, can help children cope by remaining calm and by reassuring their child. Children, especially young children, often look to their significant adults for how to react to such a crisis.

Given the scale of Hurricane Katrina, individuals living outside the primary impact area may still feel exposed to danger or may be exposed to the aftermath of the disaster. Your own children may be experiencing a reaction to this event.

Common reactions may include:

- A feeling of loss of control—feeling like they can't do anything about it.
- Self-centered reactions—Children's immediate reaction may be fear for their own safety. Children need repeated reassurance regarding their own safety.
- Different age groups may experience different reactions.

    - Preschoolers—thumb sucking, bedwetting, increased clinginess, school avoidance, withdrawal, increased conflict with parents and siblings.
    - Elementary school children—irritability, aggressiveness, clinginess, school avoidance, withdrawal, increased conflict with parents and siblings.
    - Young adolescents—sleep disturbance, loss of appetite, poor school performance, physical complaints (headache, stomachache), conflict with parents, withdrawal from friends.
    - Older adolescents—agitation, sleeping and eating disturbances, lack of energy, physical complaints, increase in acting out behavior.

As parents and caregivers there are things you can do to help your child cope.

Ways to help children cope include:

- Remaining calm and reassuring—Children take their cues from you. Acknowledge the loss and destruction, but emphasize the efforts to restore and rebuild.
- Acknowledge and normalize their feelings—Allow your child to talk about the incident focusing on their feelings and concerns.
- Monitor their television and internet viewing—It is recommended that children not have unlimited or extended viewing of the media coverage.
- Engage children in activities they enjoy—Participating in fun/enjoyable activities can foster a sense of security and "normalcy".
- Involve children in decisions about what they can do to help restore and gain back control. Activities such as fund raising and contributions to the relief efforts; planting a tree as a memorial; writing or drawing pictures for the workers; collecting canned goods and food items for contribution.

If your child is having particular difficulty with the crisis, contact your school through your child's teacher or school counselor for additional help and support.

Sincerely,

DRAFT

(Principals may edit)

Information adapted from Lazarus, P.J., Jimerson, S.R., & Brock, S.E. (2003). *Helping children after a natural disaster: Information for parents and teachers.* Available online from: www.nasponline.org

Section 6

School Crisis Response to School-Related Violence

INTRODUCTION

Schools must prevent, prepare for, and confront a continuum of violent acts, including catastrophic shootings and terrorism, homicides in the community, and more pervasive threats such as bullying. Educators have the challenge of balancing programs for school safety with efforts to create a positive learning environment (Paine & Cowan, 2009). School-based mental health providers and educators must collaboratively prepare response to school-related violence and engage in primary prevention with comprehensive planning for safe schools.

Types of School-related Violence

School shootings and acts of terrorism are usually sudden and difficult to predict, making prevention difficult and resulting in severe, long-term emotional trauma. Similarly, violence in the community can result in sudden death, injury, and severe trauma. Discussions of school violence must include community violence that impacts students.

Bullying is a different form of violence, generally not attracting headlines, but recognized as a significant threat to student wellbeing. The Alberti Center defines bullying as "a form of aggressive behavior characterized by intent to harm, repeated occurrence, and an imbalance of power between the bully and the victim." Forms of bullying include physical violence, teasing, social exclusion, peer sexual harassment, and targeting a student's race, ethnicity, religion, disability, sexual orientation or gender identity. "Cyberbullying" is aggression using emails, social media, text messages, or websites.

Prevalence

High profile violence such as school shootings is extremely rare, but prevention and preparation are essential for school safety and crisis plans. The Centers for Disease Control (2012) found that the odds of a homicide at school were 1 in 2.5 million, while the odds of being killed in the community were 1 in 21,000 (Cowan & Paine, 2013). A survey in Chicago revealed that 50% of 5th–8th grade students knew a relative who was shot at, and more than a third lost a relative or friend to homicide (Rossi & Golab, 2008).

While all forms of school and societal violence must be addressed, a focus is needed on preventing more prevalent threats such as bullying (Cowan & Paine, 2013). According to the CDC, 20% of high school students reported being bullied at school and about 15% reported being targeted electronically during the preceding year. Lesbian, gay and bisexual students reported being bullied significantly more, 34% at school and 19% online, than heterosexual peers (APA, 2016).

PREVENTION AND PREPAREDNESS

Ongoing commitment of the entire school community is required to establish safe schools (Cowan & Paine, 2013). School violence prevention strives to reduce the risk of violent behavior and minimize the effects of emotional trauma. It is imperative that schools develop plans and competence to ensure physical and psychological safety. When schools establish strong community connections, they are better able to help parents and caregivers, teachers, and students with a continuum of services to restore hope and security (Castro-Olivo et al., 2012).

Basic elements in supporting troubled or traumatized students include effective principal leadership, consistent access to school-based mental health providers, ongoing staff training and consultation, and developing family and community partnerships (Rossen & Cowan, 2013; Rossen & Hull, 2013). While appropriate safety procedures are critical, it is important to consider whether extreme security measures such as metal detectors and armed guards may interfere with the learning environment while not necessarily preventing violence. There must be a balance between building security and efforts that foster student connectedness, resiliency, learning, and social competency (Paine & Cowan, 2009; Reeves et al., 2010).

School Climate

A welcoming school climate promotes student learning and social-emotional wellbeing. Students are ready to learn when they feel safe, connected with adults and peers, and supported when they experience distress. School-wide programs encourage prosocial behavior and educational approaches to establishing rules and discipline. Students and families who feel valued and committed to keeping their school safe, support codes of conduct, conflict resolution, respect for others, and trusting relationships that encourage students to report potentially dangerous activity (Paine & Cowan, 2009). A supportive school environment involves both an awareness of strengths and a commitment to recognize and solve problems (Swearer et al., 2012).

Effective bullying and violence prevention should consider that simply mandating anti-bullying, anti-violence rules or slogans is not enough. There must be genuine student buy-in and value for prosocial behavior and connections to peers and the school community. There must be shared responsibility for the wellbeing of others, and actions that consistently support norms of mutual respect, empathy, and kindness.

Physical Safety Measures

Physical safety refers to school building vulnerability and seeks to ensure student safety by measures such as physical design, mitigation of physical hazards, policing functions, safety and crisis teams, and drills to improve preparation (Reeves et al., 2010).

Natural access control—lock all exterior doors during school hours, with only one entry point for visitors to provide easy screening. Monitor open entrances. Lock rooms and building areas when not in use (Conolly-Wilson & Reeves, 2013).

Natural surveillance—monitor staff, students, and visitors inside or outside the school building. Screen and provide an ID badge to all visitors. Escort anyone without a badge to the screening area. Natural surveillance includes school resource officers, cameras, metal detectors, and x-ray machines (Conolly-Wilson & Reeves, 2013).

School Safety and Crisis Response Teams

A key element in preventing and preparing for school-related violence is to have trained building and district level safety and crisis response teams. District or regional response is needed to support affected building teams when severely traumatic events occur. Safety teams conduct needs assessments to analyze safety initiatives, develop plans, determine strengths and needs, collaborate with community agencies, and ensure that the school staff and crisis response team receive ongoing training (Cowan & Paine, 2013).

Schools must develop their capacity for prevention and intervention through evidence-based frameworks such as the PREPaRE model (Brock et al., 2016) that facilitates flexible response and resources *before* a sudden, violent event happens. School and district crisis team response can have lasting positive effects on a school community coping with a catastrophic school shooting or terrorist attack (Castro-Olivo et al., 2012).

Strategies for Prevention and Preparedness

Suggestions for reinforcing prevention and school safety:

- Create a safe, supportive school climate with school-wide behavioral expectations, positive programs, and school-based mental health providers.

- Visible administrators and staff should welcome students and families.
- Strengthen school-family collaboration, involving school staff, mental health providers, and community agencies, including police and firefighters.
- Send letters to parents explaining safety policies and crisis prevention.
- Establish developmentally and culturally competent violence prevention programs such as anti-bullying, social-emotional learning, and conflict resolution.
- Encourage students to participate in safety planning and maintaining a safe school.
- Educate about school rules, encouraging students to report potential problems and creating systems for anonymous reporting such as hotlines or supportive websites.
- Use physical safety measures like natural access control and natural surveillance.
- Develop effective risk assessment teams and procedures developed by school-based mental health providers with input from community agencies.
- Strengthen screening and risk assessment procedures to identify and address students at risk for harming themselves or others.
- Be vigilant as perpetrators of school violence often engage in some prior behavior that causes concern and indicates a need for help.
- Recognize that perpetrators often have been persecuted or bullied themselves.
- Prevent or strictly control access to guns and weapons since perpetrators of school violence using weapons usually had easy access to them prior to attack.
- Use social media to help prevent and respond to school crises and violence.

Sources: Cowan & Paine, 2013; NASP, 2009;
Reeves et al., 2010

Suggestions for preparation and plan development:

- Schools should have crisis plans based upon a needs assessment and a model that responds to multiple types of traumatic events and levels of need.
- Establish a multitiered continuum of services for all students at the universal level, and more individualized at targeted and intensive levels according to need.
- Crisis team members should have access to ongoing training and practice drills.
- Safety and crisis teams should periodically review plans, ensuring there is an effective system of command with trained team member roles and functions.

- Train teachers and staff, and provide consultation on elements of response such as recognizing and supporting traumatized and grieving students.
- School and district teams should have a coordinated intervention plan for crises and collaborate with law enforcement and community agencies when needed.
- Schools should have multiple ways of contacting staff and parents to be informed about plans for reunification of students with primary caregivers.
- Provide sufficient access to school-based and community mental health services.
- Use data and needs assessment for planning developmentally and culturally competent interventions, and accommodating for students with special needs.
- Plan to promote protective factors and mitigate risk factors to strengthen student resilience during severely traumatic events.
- Facilitate collaborative relationships with families and school staff to aid response.
- Crises present opportunities for school leaders to examine plans, reinforce what works, address needs, and strengthen response capacity and preparedness.

Sources: Brock et al., 2016; Conolly-Wilson & Reeves, 2013; Cowan & Paine, 2013

Suggestions for bullying prevention and intervention:

- Administrators and safety teams should collect data about the extent of bullying and develop a school-wide anti-bullying policy that encourages mutual respect, social responsibility, social-emotional learning, and conflict resolution.
- Adult supervision should be improved, especially in areas identified as "hot spots" for bullying—often the cafeteria, hallway, or playground.
- Provide depression and suicide awareness training for staff and students, that encourages identification, reporting, and referral of students at risk.
- Participate in ongoing, evidence-based programs that discourage teasing and bullying, and regularly assess programs for effectiveness.
- Support evidence-based programs that encourage empathy and acceptance of diversity such as the Gay–Straight Alliance and Born This Way Foundation.
- Support collaboration, education, and consultation with school-based mental health providers, school staff, and families to confront bullying and address emotional difficulties of bullies, targets, and defenders.
- Distribute fact sheets and anti-bullying toolkits for students, parents and educators (http://gse.buffalo.edu/alberticenter.), or utilize The Empowerment Initiative (http://empowerment.unl.edu).

- Build a sense of belonging and connectedness among school students and adults.
- Use school rules and behavior management strategies in the classroom and school-wide to identify and provide consequences for bullying.
- Integrate anti-bullying awareness into the curriculum and school norms.
- Develop mentoring programs that provide positive attention for young people.
- Emphasize changes in school climate that raise awareness of bullying perpetration, and bystander and defender behavior, explicitly encouraging constructive intervention.
- Educate and support the assistance of student witnesses, who are also at emotional risk, to enact strategies to prevent and mitigate victimization.
- Provide counseling for perpetrators and targets of bullying as needed.

Sources: Jenkins et al., 2017; Novotney, 2014;
Swearer et al., 2012

RESPONSE TO SCHOOL-RELATED VIOLENCE

After catastrophic acts of violence, schools can play an essential role in providing mental health services and emotional support, restoring normal routines, and revitalizing the community (Castro-Olivo, 2012).

Cultural and Developmental Competence

Response to school-related violence should be culturally and developmentally appropriate. Before a crisis, schools can conduct a cultural inventory to understand the ways diverse groups perceive, interpret, and cope with violent events. When school personnel establish a dialogue and trust, they can better provide comfort and support. Specific functions like having trained language interpreters can be planned before a disaster. Students and families who came from countries scourged by war or terrorism may need more support. Children raised in communities where they are often exposed to homicide may live with a shattered sense of security that can interfere with their learning and emotional development, and their openness to accept support (Zenere, 2009).

Children's developmental levels affect their emotional reactions and coping mechanisms when exposed to violence. School professionals must understand how children and adolescents respond differently to crisis intervention. Younger children do not have the language skills or peer group connections that adolescents can draw upon. Children with special needs are especially vulnerable and tend to have greater difficulty coping with traumatic events and deliberate violence (Castro-Olivo, 2012).

Interventions to Support Students and the School Community

While extremely rare, violence resulting in mass casualties is profoundly traumatic and the impact long-lasting. Interventions must be broad in scope, system-wide, and long-term. After severe violence, response should address a range of anticipated reactions, reestablish perceptions of security, evaluate the continuum of individual needs, and provide services to reduce emotional trauma and further resilience and recovery.

The immediate priority when there is violence on a school campus is to get students and staff out of harm's way, request emergency help to end the danger and aid those injured, and activate the crisis plan and response team. Children look to adults for cues that can exacerbate trauma or convey a sense of calm and control. Teachers, administrators, and staff exposed to lethal violence must be emotionally supported in order to manage their stress and create a secure learning environment (Lyytinon & Palonen, 2012).

Suggestions for responding to school-related violence:

- During a violent event, follow directions from school authorities about lockdown, evacuation, or flight until law enforcement arrives to provide direction.
- Whether locked down or evacuated, enact a process to account for all students and staff, reporting those injured or missing to administration.
- Restore physical safety to the campus and security to all district schools.
- Provide psychological first aid and ongoing counseling for students, staff, and families at accessible schools and locations in the community.
- Implement procedures to reunite students with parents/guardians, ideally after providing caregiver training.
- Restore safety and perceptions of security—while most students and staff recover from trauma with natural support systems, some will need in-school support and a smaller percentage will need treatment referral.
- Provide a range of classroom, small group, and individual interventions, with groups homogeneous in terms of participants' exposure to emotional trauma.
- Conduct primary, secondary and tertiary triage, using screening tools and risk variables to determine degree of need and appropriate interventions.
- Provide support for faculty, including counseling, in-classroom support, and substitute teachers as needed, so they can better care for their students.
- Provide stress management for faculty, staff, and responding caregivers, who are at risk for vicarious trauma.
- Educate staff about effects of traumatic violence on learning and behavior—adjust coursework, offer tutoring, and provide interventions to prevent more violence, bullying, threats of suicide, and risk-taking behaviors.
- Address factors that might have led to the shooting or violent act, including bullying or gang activities.

- Social support aids recovery and strengthens resilience—provide additional recreational activities to maintain connections to school, peers, and caring adults.
- Enhance student self-efficacy and control by providing opportunities to engage in constructive, compassionate, life-affirming activities.
- Monitor social media and the internet to help identify students at risk—websites and links can be provided where students can seek support.
- Adults can help children cope by careful listening, helping them express feelings, reassuring that helpers are working to protect them, helping them problem solve, avoiding stereotypes of people or countries associated with violence, minimizing exposure to frightening media images, and being attentive to those at risk.
- Plan to set limits, but also cooperate with the media to provide useful information for families and the community.

Suggestions for planning school reentry after catastrophic violence:

- When a school has been closed after violence with casualties, student and staff reentry should be carefully planned to support perceived safety, with visible presence of administrators, teachers, sufficient school-based mental health providers, and reminders of resilience rather than trauma.
- Prior to reentry, provide caregiver training to school staff and families, including information about student reactions and managing difficult classroom situations.
- Prior to reentry the school building should be renovated so that physical reminders of a shooting or other violence are removed.
- Prior to reentry, with support of mental health providers, staff should visit the school premises to "retake control" of the building before working with students.
- Reestablishing a normal school routine is crucial in creating a sense of stability.
- School memorial activities can support recovery, but should be carefully planned with student input, should have voluntary attendance, and permanent memorials should not be placed at the school's entrance.

Sources: Castro-Olivo et al., 2012; Kennedy-Paine & Feinberg, 2014; Lyytinon & Palonen, 2012; NASP, 2015; Zenere, 2013

Suggestions for long-term follow-up after catastrophic violence:

- After severe school-related violence, recovery can take months and even years. A follow-up debriefing one to two weeks after the event should examine the team's response and needs, attend to responders' stress management, identify students who remain at risk and need treatment, and plan for long-term concerns.

- Counseling should continue to be accessible for students, staff and families, with treatment referrals when severe symptoms persist or there are suicidal thoughts.
- School staff and community should remain vigilant for signs of suicidal behavior among survivors, who may be at increased risk after a school shooting.
- While severity of reactions varies between individuals, recovery is the norm.
- Psychological and social support should be systematically planned by mental health professionals having knowledge of stress reactions and trauma.
- Emphasize violence, suicide, and bullying prevention programs, and cooperative, mutually supportive behaviors among students.
- Take into account that long-term services could create financial strain.
- Triage and screening should continue long-term, including summer months as stress reactions may be prolonged or appear spontaneously well after the event.
- Consider response to the incident anniversary date and birthdays of victims.
- Provide outreach and support for students who have dropped out, been suspended, or expelled from school.
- Consider the need for ongoing assistance for school administrators and staff.
- Crisis responders and school-based mental health providers are at risk for vicarious trauma so they should be supported and seek treatment if needed.

Sources: Brown, 2002; Lyytinen & Palonen, 2012; Rossen & Cowan, 2013; Zenere, 2013

REFERENCES

American Psychological Association (APA). (2016). Bullying prevention is a top priority. *APA Monitor on Psychology*, 16–17.

Brock, S.E., Nickerson, A.B., Louvar Reeves, M.A., Conolly, C.A., Jimerson, S.R., Persce, R.C., & Lazzaro, B.R. (2016). *School crisis prevention and intervention: The PREPaRE model* (2nd ed.). Bethesda, MD: National Association of School Psychologists.

Brown, M.B. (2002). School Violence. In S.E. Brock, P.J. Lazarus, & S.R. Jimerson (Eds.) *Best practices in school crisis prevention and intervention* (pp. 487–502). Bethesda, MD: National Association of School Psychologists.

Castro-Olivo, S., Albeg, L., & Begum, G. (2012). War and terrorism. In S.E. Brock & S.R. Jimerson (Eds.) *Best practices in school crisis prevention and intervention* (pp. 437–454; 2nd ed). Bethesda, MD: National Association of School Psychologists.

Centers for Disease Control and Prevention (CDC). (2012). *Youth violence: Facts at glance data sheet.* Retrieved from www.cdc.gov/violenceprevention

Conolly-Wilson, C. & Reeves, M. (2013). School safety and crisis planning considerations for school psychologists. *NASP Communique, 41*(6).

Cowan, K. & Paine, C. (2013). School safety: What really works. *Principal Leadership, 13*(7), 12–16.

Jenkins, L.N., Demaray, M.K., & Tennant, J. (2017). Social, emotional, and cognitive factors associated with bullying. *School Psychology Review, 46*(1), 42–64.

Kennedy-Paine, C. & Feinberg, T. (2014). Sparks Middle School: After a tragedy. *NASP Communique, 43*(2).

Lyytinen, N. & Palonen, K. (2012). Aftercare: Support for school personnel following a shooting in Finland. In C.L. Mears (Ed.) *Reclaiming school in the aftermath of trauma: Advice based on experience* (pp.135–151). New York: Palgrave Macmillan.

National Association of School Psychologists (NASP). (2009). Ten years later remembering Columbine and reinforcing school safety: Tips for school staff. *NASP Communique, 37*(6).

National Association of School Psychologists (NASP). (2015). Helping children cope with terrorism—Tips for families and educators. NASP resources. Retrieved from www.nasponline.org.

Novotney, A. (2014). An all-out anti-bullying focus. *APA Monitor on Psychology,* 63–65.

Paine, C.K. & Cowan, K.C. (2009). Remembering Columbine: School safety lessons for the future. *NASP Communique, 37*(6).

Reeves, M., Kanan, L. & Plog, A. (2010). *Comprehensive planning for safe learning environments: A school professional's guide to integrating physical and psychological safety— prevention through recovery.* New York: Routledge.

Rossen, E. & Cowan, K. (2013). The role of schools in supporting traumatized students. *Principal's Research Review, 8*(6), 1–8.

Rossen, E. & Hull, R. (Eds.) (2013). *Supporting and educating traumatized students: A guide for school-based professionals.* New York: Oxford University Press.

Rossi, R. & Golab, A. (2008). I can't go outside. *Chicago Sun Times.* Retrieved from www.suntimes.com/news/education.

Swearer, S.M., Collins, A., Fluke, S., & Strawhun, J. (2012). Preventing bullying behaviors in schools. In S.E. Brock & S.R. Jimerson (Eds.) *Best practices in school crisis prevention and intervention* (pp. 177–202; 2nd ed.). Bethesda, MD: National Association of School Psychologists.

Zenere, F. (2009). Violent loss and urban children: Understanding the impact on grieving and development. *NASP Communique, 38*(2).

Zenere, F. J. (2013). Symposium of hope: Recovery and resiliency after the Sandy Hook tragedy. *NASP Communique, 41*(7).

14 Hope and Healing

A Community Response to a School Shooting

Cathy Kennedy-Paine

INTRODUCTION

It is commonly stated, there are two types of schools in America today: those that have had a crisis, and those that are about to. The staff and students of many school communities including Thurston High School (1998), Columbine High School (1999), Virginia Polytechnic Institute (2007), Sandy Hook Elementary (2012), and Umpqua Community College (2015) all shared the belief that a crisis would not happen on their campus. That belief in the fundamental safety and security of our schools was shattered in those communities as a result of tragic shootings involving multiple victims.

As a school psychologist with 25 years of school crisis response experience, and the lead for the National Association of School Psychologists' (NASP) National Emergency Assistance Team (NEAT), I have helped guide schools across the country to recover from crisis events. I am all too familiar with the devastating effects of traumatic events, yet I have also seen how schools can help their community move forward with hope and in doing so help students and staff learn how to adaptively cope with tragedy. I have seen this in consultation work and at my own school, where I was the lead school psychologist following a tragic mass shooting. When safety is breached there is a huge effect on the school and the community, and the recovery can be long and arduous. What follows is my story, *our* story, of how one district responded to and recovered from a mass school shooting perpetrated by a student.

May 20, 1998

A freshman buys a handgun from a peer at Thurston High School, a suburban middle-class Springfield, Oregon community of 50,000 people. They exchange $100 and he stashes the gun, which his friend had stolen that morning, in his locker. Word gets out. He is suspended pending expulsion, arrested at school, and charged with possession of a loaded stolen handgun. His father is called and he is taken to the police station—both are interviewed. His father is a teacher at the local community college, having recently retired from teaching at Thurston High. The father denies (falsely) that they have any weapons at home. With no reason to suspect imminent danger and no prior arrest record, the police release the student according to the law at the time. They drive to their rural Springfield home.

The student's father desperately calls a residential treatment facility several hundred miles away, but is told his son is too young for their program. At that moment his son silently enters the kitchen. With his father's back to him he raises a gun his father had bought for him, fires, and kills him. Later, his mother returns home after teaching high school Spanish. A popular Springfield High School teacher, she has just been named "Outstanding Educator of the Year," but will not receive her award. As she gets out of her car in the garage, her 15 year old son appears in the doorway, says, "I love you, mom" then shoots and kills her. He spends the night setting explosives throughout the house.

May 21, 1998

The day begins like any other at Thurston High School with students eating breakfast in the cafeteria and trading tales of youthful innocence. At 7:56 a.m. the freshman enters the back of the school near the tennis courts with four weapons concealed under his tan trench coat—a .22 semi-automatic Ruger rifle, a 9mm Glock pistol, a .22 Ruger MK II pistol and a hunting knife strapped to his leg. He is carrying 1,100 rounds of ammunition in a backpack. He tells several students to leave, then shoots two students as he walks down a hallway. He enters the cafeteria, pulls the semi-automatic rifle from beneath his trench coat and randomly sprays 50 rounds of ammunition into the 300 students gathered for breakfast. What is first thought to be a prank soon turns into a nightmare. When he stops to reload his rifle one of the injured students closest to him recognizes the pause. He lunges at him and instantly is joined by six other students tackling the assailant to the ground. The school goes into lockdown—staff members call 911 and begin administering first aid to the wounded.

Soon it is a chaotic scene of parents, reporters, police, and emergency workers all trying to learn what happened, trying to find their children. A dazed teacher directs traffic, the same teacher who, moments before, had administered first aid to wounded and dying students. Many emergency responders perform their duties without knowing the fate of their own children who attend the school. Parents fill the sidewalks and press past the gathering media to reach the school. We are the fifth major school shooting that school year in the United States.

As a district psychologist and Crisis Team Leader, I receive the emergency call just a few minutes after the shooting. I rush to the school passing one ambulance after another as they race by with victims. Twenty-two students are transported to area hospitals in eleven ambulances—one ambulance leaves the school every four minutes for forty-five minutes. In the end, two students are killed and twenty-five wounded. We are transformed from innocent individuals starting a normal day, to traumatized victims of a school-shooting spree. Principal Larry Bentz sees shock, disbelief and tears on the faces before him as he reads the names of the wounded to the gathered parents. Time stands still. Our world is filled with sirens, ambulances, stretchers, reporters, police cars, yellow tape, flashing lights, and sobbing voices.

An eerie quiet prevails inside the school. Three hundred students who witnessed the shooting and survived gather in the library where caring adults calm them as they await police questioning. Frantic parents search for sons and daughters, some who will never come home. There have been many near misses and close calls. One student removes his backpack and discovers a bullet lodged in the middle of his history book. Images I still carry in my mind: police, counselors, blood stains, darkened rooms, students huddled, phones ringing, backpacks strewn, quiet sobs, parents searching, anguished looks. The student attacker is taken to the police station for the second time in two days—this time with much more serious consequences.

PREPAREDNESS AND DESCRIPTION OF RESPONSE

When President Clinton phoned the afternoon of the shooting, we realized that this tragedy would affect not only the 12,000 students and employees of the Springfield School District, but the entire community, and the nation. Our sense of safety and security was shattered along with our innocence. We also learned, on the ground, what worked and what did not in terms of our prevention and response efforts, and what it took to recover from a serious traumatic event.

Homicide Impact

The broad issues that made this crisis difficult were the nature of the crime—homicide—that destroyed the safety and security of our school, and the pervasive, community-wide impact of the event. Stephen Brock and colleagues noted, "Generally, human-caused events, particularly those that involve personal assault by someone who is familiar to the victim, are more distressing than natural disasters" (Brock et al., 2016). The issues in supporting homicide survivors are complex. Debra Alexander, a trauma consultant stated, "Grieving the loss of someone who has been killed suddenly, violently, and senselessly is different from any other form of grieving. The anguish is intense and long lasting. The physical and emotional reactions to the trauma are only the beginning. Criminal justice systems, insurance companies, settlements, and media can present a multitude of frustrations and often repeatedly cause a return to initial trauma reactions" (Alexander, 1999).

The impact of this trauma blanketed our community. In addition to the shooting victims, so many others were affected—students and staff who were in the courtyard and in the adjacent hallways; the parents, family, friends, and classmates of the critically wounded; the deceased parents' friends and teaching colleagues at Thurston High, Springfield High and Lane Community College; secretaries, custodians, coaches, cafeteria workers.

Crisis Models

Schools must be prepared to respond to a variety of crisis events. School crisis events are extremely negative, uncontrollable, and unpredictable (Brock et al., 2009) and they produce a variety of victim reactions that can result in severe mental health challenges. Some of these reactions such as depressive disorders, anxiety disorders, trauma-related disorders, dissociative disorders, sleep-wake disorders and substance-related and addictive disorders can appear in school settings and affect learning (Brock et al., 2009). Because of these potentially negative effects, schools must be prepared to address all phases of a crisis: prevention, protection, mitigation, response and recovery (U.S. Department of Education, 2013).

Crisis response models have been developed over the years for use in schools, many of them since our tragedy. In 1998 we used the available models, including those developed by Jeffrey Mitchell (Mitchell, 1983; Mitchell & Everly, 1996) and Robert Pynoos (Pynoos et al., 1995). The PREPaRE model is a comprehensive curriculum of school crisis prevention and intervention developed for school-based professionals that provides training on the roles and responsibilities of safety and crisis teams (Brock et al., 2009, 2016). Although the model did not exist at the time of our crisis, in hindsight, we used many of the strategies it advocates.

Prior Crisis Planning

In an ideal world, we would never need a crisis plan to respond to a school shooting. In the real world, we do. The first element of the PREPaRE model is crisis prevention and preparedness. Two important factors that aided our recovery were the planning we had done prior to the event and the community relationships we had established. While nothing in our previous experiences with individual student and teacher deaths truly prepared us for the magnitude of this horrifying event, one fact is clear—if we had not had a crisis response plan and a trained team we would not have been able to effectively respond to this tragedy. Within an hour of the shooting my colleagues and I organized a core team of school psychologists, school counselors, administrators, and community mental health workers. Together we designed the school district's mental health response—an on-the-spot modification of procedures we'd used in dozens of crisis interventions over the previous seven years. It was reassuring to know that our basic plan was sound and we just needed to expand and adapt it to the situation. At the same time, a multi-disciplinary team of district-level staff and emergency responders orchestrated the rest of the response using a Unified Incident Command System (FEMA, 1995). Although district staff had not been formally trained in the Incident Command System, our Superintendent was experienced in ICS and directed us during the response.

At the time of the Thurston shooting each school had an annually updated crisis plan, which specified the duties of team members and procedures to follow in emergencies (Paine, 1994). District counselors and psychologists were trained and experienced in crisis response and a county network of trained crisis personnel was in place. Immediately after news of the shooting was broadcast, counselors and mental health workers responded to Thurston High. Our network allowed us access to over 100 counselors from other school districts and from our local mental health agencies. Along with law enforcement, these counselors helped us to "reaffirm physical health and security, and perceptions of safety," and begin to "evaluate psychological trauma," the second and third steps of the PREPaRE model (Brock et al., 2009, 2016).

Challenges in Our Response

School crisis response is complex, and we were presented with five major challenges during this response. We were able to meet these challenges and lessen the impact of the long-term effects by following our plan and being flexible when things did not go as expected. Here are the challenges we faced and the critical aspects of our response:

(a) Coordinating the School District and Responder Agencies

It quickly became clear that this was a multi-jurisdictional response that required many cooperative decisions by the District Superintendent, City Manager, Mayor, Springfield Police and Fire Chiefs, County Sheriff, Oregon State Police Chief and hospital administrators. School administrators and emergency responders had previously collaborated in drafting the district's Emergency Procedures Manual (Paine, 1994), and a mock drill had been held at Thurston High with local emergency responders. As a result of that drill they knew the layout of the campus, how to get in and out efficiently, and who to contact. The challenge for the heads of these many agencies was to work together in ways they had never done before.

In order to avoid conflicting messages, school administrators and city officials quickly formed a command center at City Hall. This became a clearing-house for inquiries from both the press and the public. Press conferences were held there daily and additional phone lines set up by 10:00 a.m. the day of the tragedy were staffed 24 hours a day through the four-day holiday weekend by city and school district employees providing information to the community. To ensure coordination, the CEO's from the district, city, and the four law enforcement agencies worked together in a Unified Incident Command structure. This multi-agency team spent hundreds of hours together developing and revising plans for the schools and the community in the weeks following the shooting.

(b) Handling the Intense and Intrusive Media

Responding to the crush of local and national media was a challenge we had not faced prior to this event. Throughout the first day and night the media vans and satellite trucks rolled in from across the nation. Before the first hour had passed, a CNN helicopter hovered overhead, transmitting images of our newfound horrific "fame." Reporters from as far away as Japan, Portugal, England and Australia quickly took on a larger-than-life presence in our normally quiet community. ABC, NBC, CBS, NPR, PBS, *Inside Edition*, *Hard Copy*, *USA Today*, *Time*, *Life*, *Newsweek*, *People*, *Rolling Stone*, *and Psychology Today* made appearances. Before long, a surrealistic scene developed. The street in front of the high school was reduced to a one-lane road, with cars forced to crawl between the constantly humming generators and blazing lights of 20 white satellite vans. The city ran additional power and internet lines above the road to facilitate communication for all media.

Our district Public Information Officer and Superintendent handled the majority of the communication with the media through regular press conferences so the hundreds of reporters covering the event could get information from one source. The press conferences were scheduled on a regular basis and always included representatives from the school district, law enforcement and the two area hospitals that had received the victims. While local media were helpful and respectful, we were amazed to observe that some national reporters tried posing as doctors or counselors in their efforts to access hospitals and schools; ID badges became essential and were quickly printed for all volunteers.

Local and national television crews filmed live reports in front of the school daily, but no media personnel were allowed on the high school campus until after school resumed, and then only briefly when the damage had been repaired. Eager to film the crime scene, media members crowded into the repaired cafeteria late one night, after all students and staff were gone. This was done to minimize the filming of traumatic images and to allow the students' first view of campus to be in person, not on television. The only pictures taken of the cafeteria rest in the hands of law enforcement and have not been shown to the public. As noted by Brock, "minimizing exposure to the crisis event itself, its immediate aftermath (including the suffering of others), and subsequent media coverage may also foster a sense of psychological safety" (Brock et al., 2016). In addition, we kept the media off the campus and encouraged students and staff not to talk to them. As the days passed students posted hand-written signs saying, "No Media!" Six days following the shooting, the fleet of white vans crept silently away almost as abruptly as they had arrived, leaving the school free of lights, cameras, and sound bites.

(c) Maintaining Communication

Immediately after emergency responders arrived, we were challenged with the jamming of the district landline phone system. This prevented communication to and from the high school, and with the other district schools. In addition, there were many rumors and at times it was difficult to determine when information was credible. We used newly acquired administrative cell phones, pagers and our district two-way radios for communication. I grabbed a cell phone when the landlines went down in order to call for additional counselors.

A church across the street was immediately opened and parents were directed there to receive information about injured students. Unfortunately, there was a delay of what seemed like days—really less than an hour, before the police could release the names of the victims. At noon that day we met with the Thurston High School staff. We had just begun to absorb the news of the student deaths and injuries when we were told of the assailant's parents' deaths. Many in the room were close friends with them and the shock was audible. We sent the staff home. Later in the day we briefed the central office staff and all the district administrators from our 23 schools. By 4:00 PM approximately 125 counselors gathered at the school from the community, offering their support. Screening and organizing those counselors became a major task for us.

Communication with our district schools and with other schools in the area was also a challenge. We are a tightly knit community—staff at Thurston had students in other schools, and staff in other schools had students at Thurston. In hindsight, we could have used the district email system and website to relay information to other schools and the public more effectively.

(d) Planning the Students' Delayed Return to School

The high school was closed for four days while police gathered evidence and repairs were made. The shooting had occurred just prior to the Memorial holiday weekend. Our challenge was to provide mental health support to students, staff and parents during this period, and to support them in returning to the campus the following week. The importance of mental health interventions is detailed in step four of the PREPaRE model, "provide interventions and respond to psychological needs" (Brock et al., 2009, 2016). The Red Cross provided meals, and 80 trained therapists, including school-based mental health providers from our district, county school districts, community and private practice counselors, met with students, teachers, parents, and administrators the day following the tragedy in a nearby middle school that was closed for the day. Because the slain mother taught at Springfield High, those staff and students were severely impacted and required support. Our district counselors and psychologists counseled students and staff through the weekend, and managed the monumental task of screening, scheduling and monitoring ultimately over 200 outside counselors. Volunteers from the National Organization of Victims' Assistance (NOVA) and the National Association of School Psychologists' National Emergency Assistance Team (NEAT), who arrived the day after the shooting, held community debriefings, news conferences, and training sessions throughout the next nine days using the NOVA model (Young, 1998).

An important step of the recovery occurred when students, their families, and staff visited the repaired Thurston campus during an Open House on May 25th, Memorial Day. This allowed everyone to enter the campus supported by family, friends, counselors, and even "comfort dogs." Although many of the 2,000 visitors sat or stood in the repaired cafeteria, not all were able to be there. Many tears were shed that day and one student commented:

> It was unlike anything I've ever felt before. I've been around death before in my life but nothing like this . . . nothing that's really just gotten down to the very core of me and made me want to break down and cry right there.
>
> (Anonymous student interview, May 25, 1998 at
> Thurston High School)

The final step of the students' reentry occurred the first day of classes, the week following the shooting. After a free breakfast in the school court-yard and cafeteria, students attended a half-day of classes. Hundreds attended the funeral of one victim that afternoon. Volunteer counselors were present in every classroom that day, and remained available in support rooms through the end of the school year. The focus of those first few days and weeks was to assist people in coping with the trauma, while returning to routine. During the final three weeks of school, we maintained a regular schedule of classes, with trauma support rooms continually staffed by our community mental health volunteers. We greatly appreciated all of this support, as the impact of the trauma was pervasive throughout our community, and we were overwhelmed with the tasks at hand. Personally, I found the work to be both physically and emotionally draining. My work on the hundreds of organizational details was constantly interrupted by newly emerging issues or problems. My days were spent at the school and my nights preparing for the next day.

(e) Supporting the Long Road to Healing

The long-term recovery from this trauma required complex support from many sources, including private counseling agencies, Lane County Mental Health, Lane Education Service District, other school districts, NOVA, NEAT, and the Red Cross.

RECOVERY

Following are some of the important factors that were part of our long-term recovery process:

Time, Staff and Additional Resources

The event was not over on May 21st, the day of the shooting, or on June 12th, the last day of school. Summer was filled with grant writing, activities at Thurston High, planning for freshman orientation and the first days of school, and training district staff in the dynamics of post-trauma reactions. The Thurston Assistance Center was established to provide counseling support and information to Springfield students and families affected by the shooting. Funding is often an issue for district response and since the Columbine tragedy, the federally funded "Project SERV" (School Emergency Response to Violence) grants have proven very useful to schools recovering from the trauma of school violence (U.S. Department of Education).

In July, many of us attended a symposium led by Robert Pynoos, a premier expert in the field of trauma. We were privileged to have Dr. Pynoos and his colleagues return to Springfield to work with district administrators, school board members, teachers and staff members in post-trauma responses. Marlene Young, Executive Director of NOVA, conducted a three-day crisis response workshop for 50 counselors, psychologists and mental health workers in the community.

Return to School in the Fall

Springfield Superintendent Jamon Kent and the school board members made a commitment to enter the following school year with thoughtfulness, planning and training in response to this tragedy. Our approach was designed to recapture the school's normal activities, and at the same time reaffirm perceptions of safety and security, and facilitate a healthy recovery for students and staff. We added uniformed School Resource Officers to both high school campuses. The reentry to Thurston High was carefully planned and supported, beginning with the Memorial Day open house the previous spring. The cafeteria was painted and brightened to minimize traumatic reminders, yet we knew there would be many reminders ahead. Most students and staff were able to cope, with the help of their naturally occurring support systems. For those more severely impacted we provided a combination of psychological education, group crisis intervention and individual crisis intervention, tenets of the PREPaRE model (Brock et al., 2009, 2016).

On the first day of school in the fall, reporters and media trucks surrounded the school once again as students entered the campus filled with much excitement and some apprehension. Teachers asked students to be tolerant and patient with one another as they worked through a broad range of reactions, and reminded them that while many students were ready to move on, some were not. Of the 25 injured students, 21 returned to Thurston High. Some still carried the physical evidence of scars and bullets within them and faced lengthy rehabilitation. Some could not yet return to the cafeteria and feared recurring violence. Bereavement was complicated by traumatic grief. The prevailing atmosphere that day, however, was reflected by the words of one senior:

> Though we were inevitably affected by tragedy, we are looking forward to what life has to offer us next. We have learned how very precious, yet circumstantial, life is. Now, more than ever, our eyes are open wide, our ambitions are high, and we are ready to live.
>
> (Anonymous student interview, September 1998 at Thurston High School)

In October, we continued to evaluate the level of psychological trauma. With the help of Dr. Pynoos, we conducted a trauma screening for all the students at Thurston High by adapting his trauma screening index (Pynoos et al., 1998). We added two additional high school counselors through grant funds to provide group and individual counseling for staff and students until all students present at the time of the shooting graduated three years later.

Care of the Caregivers

The two NOVA teams held numerous debriefings that were essential for the emotional wellbeing of the service providers. They supported us with wisdom and caring during the most dreadful period we could imagine. We all experienced *compassion fatigue* and *secondary traumatic stress* (Figley, 1995). Personally, I was unprepared for the emotional toll this would take. Bolnick and Brock (2005) conducted a study to document the effects on school psychologists of providing mental health crisis interventions. The most commonly reported reactions included fatigue and exhaustion, increased sensitivity, anxiety, difficulty concentrating, helplessness, sleep difficulty, irritability and preoccupation with the event. Yes, I felt all of these and ultimately sought my own personal counseling support.

I also experienced a phenomenon that Brock and colleagues described in *School Crisis Prevention and Intervention: The PREPaRE Model*: "School-employed mental health professionals who provide crisis intervention services . . . have the dual challenge of providing mental health crisis intervention while also fulfilling other demanding job responsibilities and maintaining existing caseloads" (Brock et al., 2016). My job as a school psychologist and district special education administrator continued, even as I took on the responsibility for planning the mental health recovery of students and staff.

Memorials and Healing Events

The most impromptu of the memorials became one of the most powerful for a community looking for solace in this tragedy. Within hours of the shooting, community members of all ages placed flowers, posters, balloons, plants, teddy bears, candles, photos, poems, crosses and other mementos on the chain link fence in front of Thurston High. Ultimately, this memorial stretched the entire length of the campus, some 150 yards representing the community's outpouring of grief in a sea of flowers and gifts. For several days, vehicle and pedestrian traffic was non-stop, as thousands passed to pay their respects. In addition, there was a candlelight vigil outside City Hall, a memorial service for the slain parents, prayer services in many churches, and the firefighters' Blue Ribbon of Promise campaign to end school violence began (Blue Ribbon of Promise, 1998).

The design and construction of a permanent memorial proved to be one of the largest and most complex challenges of our recovery. An initial committee of students, staff, and parents designed a permanent school memorial. The process stalled due to lack of funds and lack of agreement about the details of the design. Several years later a new group was formed and the memorial was finally dedicated May 21, 2003, on the fifth anniversary of the shooting. The memorial consists of a small park near the high school containing trees, benches, a representation of the Thurston memorial fence, and a basalt pillar. The engraving on the pillar honors those who lost their lives, those who were injured, and those who came to their aid. It states in part:

THIS MEMORIAL SHALL STAND FOREVER IN MEMORY OF Mikael Nickolauson and Ben Walker. Appreciation is extended to the community for opening their hearts and offering help in so many ways; from assistance provided by local medical facilities and personnel to donations and continued support. The courage and strength shown by the victims and families has inspired all, and has given hope and encouragement to continue with life after tragedy. May we all understand the life changing impact of violence, and may this place extend the comfort, strength and hope that comes from a caring community, state and nation.

One-Year Anniversary

The days leading up to the one-year anniversary were filled with anticipation, anxiety, rumors of copycat violence, and daily doses of media attention (all intensified by the Columbine shooting one month prior to our anniversary). The day began with bomb-sniffing dogs searching the campus, dozens of parents patrolling the area, and continued with school and community counselors, and police officers supporting and reassuring the two thirds of the student population who attended classes. While most had returned to a normal routine, some adults and students experienced a resurfacing of emotions. We saw a broad range of reactions of varying intensity depending on the individuals' personal history and relationship to events. Many students exhibited little to no effects, while others re-experienced anxiety, fear, anger, or grief like the emotions felt during the tragedy. The day ended as over a thousand people gathered to remember the families of the victims in a "Community Gathering for Remembrance and Renewal" at Thurston High School. Teresa, our most seriously wounded student addressed the crowd via video. Shot in the head and given just a 10% chance of survival, she made a remarkable recovery, despite life-long challenges. Her simple plea: "Please. . .stop this violence."

The Legal Process

Our community had previously endured high profile murder trials, but we had never experienced anything that attracted the attention of the world. In preparation for the trial we sought the advice of experts and held multiple planning meetings involving the staff of the school district, City of Springfield, District Attorney's office, victim's assistance workers, public information directors and trauma counselors. Debra Alexander has described the unique dynamics of homicide and the legal system. She said:

> Homicide creates a different kind of grief because of the rage it evokes. Often the criminal justice system slows down the grieving process when closure around a case is not made or appears to be made unjustly. Children may need your help as they struggle to understand the act of murder and the intentional taking of a human life. They will need your continued support and understanding as the legal case unfolds.
>
> (Alexander, 1999)

As we struggled through our grief and anger we tried desperately to understand how a 15 year old boy could commit actions that caused so much harm. He grew up in a middle class family with his older sister and two parents who were teachers. In a media interview, the sister acknowledged that her brother sometimes had a difficult time at school. However, she added, "Until that day, there was nothing that could make us believe that something of this scale was possible." A PBS *Frontline* documentary described the family context as "a nurturing home, a comforting community, and loving parents recognized for their special way with children—by all accounts an ideal American family" (Lieberman, 2006).

We do know that in middle school this troubled youngster began seeing a psychologist because of his mother's concern about his hot temper and fascination with explosives, guns and knives. He told the psychologist that he often felt angry, without knowing why, and set off explosives to vent those feelings. The psychologist diagnosed a major depressive disorder. By July 1997 he no longer seemed depressed and had stopped setting off explosives, so his counseling and medication were discontinued. At about that time, at the youth's urging, his father purchased a handgun and a semi-automatic rifle for him with the agreement that he would not use the guns without his father present. Unknown to his parents, he also began obtaining guns from friends and downloading bomb-making instructions from the internet (Lieberman, 2006).

The seriousness of the youth's complex mental health condition was largely underestimated at the time. During the sentencing hearing psychologists testified that, upon evaluation, he suffered from paranoid schizophrenia. Following the assailant's arrest, he told psychologists that voices in his head told him to shoot his parents and the students at Thurston. He said he had been hearing voices since age 12, but never told anyone about the voices because he was afraid of them. A handwritten note found in his family's home the morning of May 21, 1998 said, in part:

> I have just killed my parents! I don't know what is happening. I love my mom and dad. I just got two felonies on my record. My parents can't take that! I am a horrible son . . . my head just doesn't work right . . . God damn these VOICES inside my head. I want to die . . . But I have to kill people. I don't know why. I have never been happy . . . I have no other choice. I am so sorry.
>
> (Lieberman, 2006)

Eighteen months after the shooting he pled guilty to the charges just four days before jury selection was to begin, avoiding a trial. During a seven-day sentencing hearing multiple victims spoke of the impact of this tragedy in emotional, heart-wrenching testimony reflecting anger, hate, sadness and sorrow. On November 9, 1999, under the state's mandatory sentencing law, he was sentenced to serve 112 years in prison, without parole.

Moving Forward

The mental health recovery of the students and staff of Springfield is an on-going process. Even today we are sometimes reminded of this tragedy's impact on our community. We now have district employees and parents of current students who themselves were students at Thurston during the shooting. We are more cognizant of the elements of school safety, and School Resource Officers continue to support our schools. In 2008 our county received a federal "Safe Schools Healthy Students" grant that provided funds for all districts to train staff using evidence-based programs in school crisis response. The project was a joint collaboration of three school districts and the University of Oregon's "Institute on Violence and Destructive Behavior" and included the components of Crime Prevention Through Environmental Design, Effective Behavior Support Plans, Crime Prevention Specialists, School Mental Health Specialists and the Turnaround School for students facing expulsion for violence or threats of violence.

As a result of this tragedy, our community came together as never before. We can't go back to the way we were. Our task has been to integrate this trauma into our lives, to move on and find a "new normal" for each of us. As Principal Larry Bentz stated:

> The tragedy cannot remain the story. The story has to develop into the character of the people who were involved, and the strength and the bravery and the courage that will pull us forward, because *hope is what leads the way.*
>
> (Bentz, 2006)

LESSONS FOR ALL

The fifth and final element of the PREPaRE model focuses on evaluating school safety and crisis response strategies (Brock et al., 2009, 2016). As we look back and examine the effectiveness of our response we feel confident that the strategies we used were solid and match the best practices being advocated today. As we think about the supports needed to make sure our schools are safe places, we believe that three strategies are critical for all schools:

(1) Schools Need Systems and Trained Staff to Immediately Respond to a Variety of Crises

Because we cannot prevent all crises from occurring, it is clear that every school must be prepared with a crisis response plan and trained personnel. A comprehensive plan that includes crisis preparedness and response procedures is the foundation of a safe school. Schools must form a crisis response team, develop a comprehensive written crisis response plan that covers a range of possible events, coordinate the plan with local emergency responders, conduct training of all school staff, and integrate the elements into a comprehensive school safety plan. The effectiveness of recovery is extremely dependent on the quality of prior planning and preparation as well as the character and connectedness of the school climate. The school's capacity to follow through with steps toward recovery is shaped by the school's preparedness and culture. As we learned at Thurston High, having a district crisis plan and training allowed us to quickly take action and begin supporting the students and staff. Annual training and plan evaluation is necessary and requires continuous, sustainable funding for preparedness.

(2) Schools Need Adequate and Sustainable School-Based Mental Health Services

The perpetrator of our school shooting in Springfield was diagnosed with mental illness subsequent to his arrest. Although his illness was ultimately beyond the scope of a school intervention, school counselors, school psychologists, and school social workers can provide critical services to meet a range of mental health needs in conjunction with community resources. School-based services that are linked to learning also improve academic performance and problem-solving skills. Providing quality, comprehensive services for students beginning in the early years requires that schools have an adequate number of appropriately trained staff, as well as close collaboration with community providers.

(3) Schools Need a Plan for Students at Risk of Violence

It is critical to reinforce the importance of reporting concerns about young people hurting themselves or others, and enhance school connectedness and trust between students and adults. We know that reporting is one of our most effective preventive measures (Fein et al., 2004). Over the years, our perpetrator made occasional threatening comments to peers, his parents, psychologists and teachers. When questioned, he always said he was "just joking". There was no process where all the individual concerns could be viewed together. And so, in collaboration with school mental health professionals, each school should develop a plan to identify and support those students at risk for violence to themselves or others, or who are experiencing mental health difficulties. It is important for schools to coordinate with community resources and have the staff and resources for early identification and referral of potentially violent youth. This includes a formal threat assessment team including a trained mental health professional, school administrator and safety officer. A promising model that we adopted, was developed by forensic clinical psychologist, Dewey Cornell (Cornell, 2010).

The lessons learned in Springfield may be applied to any major traumatic event in a community. It is no longer a question of whether a crisis will strike; it is a matter of when. The time to prepare is now.

REFERENCES

Alexander, D.W. (1999). *Children Changed by Trauma: A Healing Guide*. Oakland, CA: New Harbinger Publications.

Bentz, L. (2006). *How a Principal and Town Cope with a Shooting*. National Public Radio. Retrieved from: www.npr.org/templates/story/story.php?storyId=6211225& from=mobile

Bolnik, L. & Brock, S.E. (2005). The self-reported effects of crisis intervention work on school psychologists. *California School Psychologist*, 10, 117–124.

Brock, S.E., Nickerson, A.B., Reeves, M.A., Jimerson, S.R., Lieberman, R., & Feinberg, T.A. (2009). *School crisis prevention and intervention: The PREPaRE model*. Bethesda, MD: National Association of School Psychologists.

Brock, S.E., Nickerson, A.B., Reeves, M.A., Conolly, C.N, Jimerson, S.R., Pesce, R.C., & Lazzaro, B.R. (2016). *School crisis prevention and intervention: The PREPaRE model, second edition*. Bethesda, MD: National Association of School Psychologists.

Cornell, D. (2010). *The Virginia student threat assessment guidelines*. Author: University of Virginia. Retrieved from: http://curry.virginia.edu/research/projects/threat-assessment

Federal Emergency Management Agency (FEMA). (1995). *Incident Command System Self-Study Unit*. Jessup, MD.

Fein, R.A., Vossekuil, F., Pollack, W.S., Borum R., Modzeleski, W., & Reddy, M. (2004). *Threat assessment in schools: A guide to managing threatening situations and to creating safe school climates*. Washington, DC: U.S. Secret Service and U.S. Department of Education. Retrieved from: www.2ed.gov/admins/lead/safety/threat assessmentguide.pdf

Figley, C.R. (1995). *Compassion fatigue: Coping with secondary traumatic stress disorder*. New York: Brunner/Mazel.

Lieberman, J. (2006). *The shooting game: The making of school shooters*. Santa Ana, CA: Seven Locks Press.

Mitchell, J. (1983). "When Disaster Strikes: The Critical Incident Stress Debriefing Process." *Journal of Emergency Medical Services*, 8, 36–39.

Mitchell, J. & Everly, Jr., G.S. (1996). *Critical incident stress debriefing*. Ellicott City, MD: Chevron.

Paine, C. (1994). *Administrator's guide to crisis response*. Springfield, OR: Springfield School District.

Pynoos, R.S., Steinberg, A.M., & Wraith, R. (1995). A developmental model of childhood traumatic stress. In D. Cicchetti & D.J. Cohen (Eds.), *Manual of developmental psychopathology*, Vol. 2, 72–95. New York: Wiley.

Pynoos, R., Rodrigues, N., Steinberg, A., Stuber, M., & Frederick, C. (1998). *The UCLS PTSD Reaction Index for DSM*. Los Angeles, CA: UCLA Trauma Psychiatric Program.

Ribbon of Promise National Campaign to End School Violence. (1998). Retrieved from: www.rwjf.org/portfolios/resources/grantsreport.jsp?filename=035695.htm&iaid=141

U.S. Department of Education (2013, June). *Guide for developing high-quality school emergency operations plans*. Washington, DC: Author. Retrieved from: http://rems.ed. gov/docs/REMS_K-12_Guide_508.pdf

U.S. Department of Education Project School Emergency Response to Violence (SERV). Retrieved from: www2.ed.gov/programs/dvppserv/index.html

Young, M. (1998). *The Community Crisis Response Team Training Manual*. Washington DC: National Organization for Victim Assistance.

15 Responding to Shootings in My School District

Commitment to Crisis Prevention and Intervention

Scott Poland

INTRODUCTION

This narrative provides an overview of three school shootings that occurred from 1985 to 1992 in the Cypress Fairbanks Independent School District in Houston, Texas, where I worked as the Director of Psychological Services. It describes my personal response and the improved district crisis prevention and intervention resulting from these tragic shootings. The third shooting was the only one that resulted in a death. The victim was a staff member who I knew personally very well. I will recount the initial reactions to each shooting and the efforts to assist the affected staff and students. Additionally, the narrative describes the creation of building crisis teams with designated liaisons, and how the planning and training resulted in a comprehensive team response to the third shooting.

UNPREPARED FOR RESPONSE

Response to a Shooting at Millsap Elementary School

I was about to go into a special education meeting in 1985. The meeting was one in which there was a lot of contention about services. The parents had brought an advocate, and my boss, who was then the assistant superintendent, was present at the meeting.

We were both shocked to learn, just before the meeting, that an eleven year old boy had been shot while raising the flag at Millsap Elementary School and that students had been fired on at two other elementary school locations. I will tell you quite honestly, aside from my disbelief, my first reaction was, "What about my son?" My son didn't go to Millsap Elementary, but he was a student at a nearby elementary school. To be honest, what I wanted to do more than anything was to go get Jeremy, take him home, and make sure he was safe! Then I thought his principal would be doing the right things to keep him safe and my responsibility was to go immediately to Millsap Elementary School and do something to help. Millsap was actually my school, where I personally worked instead of sending in an intern. It was my school!

I remember gathering my books, papers, and notes to leave the special education meeting. My boss turned to me and asked, "Where are you going?" I responded, "To Millsap Elementary School, of course!" She said, "No, you're not! We're having a special education meeting here. You need to sit back down!" I am embarrassed to tell you that I did sit back down to several hours of a meeting. As soon as the meeting was over, I rushed to the nearest phone and I tried to call Millsap. I could not get through.

I had no idea really about what to do at the scene of a school shooting . . . what my role should be. I had never received any training in crisis intervention. But, I knew that I should respond somehow. My first thought was to go to my office for just a moment . . . it's nearby . . . let me sit behind my desk for a moment, and maybe I will get some miraculous idea of what I should do to help out.

When I arrived at my office, the phone immediately rang. It was CBS television in Houston, Texas. They had a simple question. "Dr. Poland, what do parents in the Cypress Fairbanks school district need to do to help their children in the aftermath of this tragic shooting? And, we want to do an interview with you!" I had never done an interview in my life. I had no idea what the media policies were for the school district. But I was pretty quick thinking! I responded, "Sir, I am sure the superintendent wants to be the single spokesperson on this issue and here is the number of the superintendent's office." I thought I handled that pretty well, but my phone rang five minutes later. It was the superintendent's office. "Scott, we believe in being cooperative with the media. We believe it would be a good idea if you give that interview. It will be helpful to our families, our students and their parents." A few minutes later, the mobile television crew pulled up in front of my building. One thing that I came up with bears repeating all these years later . . . you see, I knew what the parents were thinking, including myself—the world was a pretty awful place to have an eleven year old boy shot raising a flag in front of his elementary school. I said, "Parents, pay attention to your child's emotions. Please give them permission for a range of emotions and do not project all your thoughts and worries about the world onto your child." That was the first interview I had ever done. Today, having done as many as a thousand, I have gained confidence about interviews.

I believe that school psychologists have a lot of important information and we should get that information across by all means possible, including television, radio, newspapers, and social media. I want to clarify that the boy who was shot raising the flag recovered. The President of the United States phoned him that night while he was in the hospital. The gunman, however, was on the loose. The next morning I remember talking to the staff at Millsap Elementary School . . . the gunman had not been found and we had not closed school. There was a sense of hysteria about a gunman on the loose. I remember the bedraggled look of the teachers because they had not slept well. I said to them, "I bet your students slept better than you did. Working together, we are going to get through the day, help each other, and then we'll be in a better position to help our students." Later that day, the gunman turned himself in. All he ever stated was that he was mad at the world and that children and schools were an easy target. The child who was shot made a full recovery, and I have to say that six months after that shooting everybody was thinking, "Did that really happen?" "Yes, it really happened! But nothing like that could ever happen again in our top rated suburban school district."

First Shooting at Langham Creek High School

It was one year after the Millsap shooting, a Friday, which was a "sacred day" for my department. No one was assigned to a school on Friday. You might go to a school because there was some kind of an emergency or to catch up with an assessment or counseling. Friday was a day to be in the centralized psychological services offices for collaboration, supervision, and meetings. We were having lunch when the secretary rushed in and said there had been a shooting at Langham Creek High School. My first thought was incredible denial! "How could that have happened? That school just opened! It has only half of the student body so far, and the school cost more than 40 million dollars!" Then I was told that the assistant principal was shot in the cafeteria in front of more than 600 students and colleagues. Now, I'm angry! "How could anybody have shot him at our school?" My secretary then said, "They called and they want YOU immediately!" Now you can forget the denial and anger! Instead, think of panic and anxiety! I remember having to say to myself, "You can stay calm, you can be collaborative, you've helped people before, and you can do it again." I was giving myself a pep talk. I even had to be as specific as, "Scott, turn the wheel to the right. Langham Creek is west. You will never get there if you don't turn and go to the right."

When I pulled up to Langham Creek High School, "life flight" was just taking off with the wounded administrator. If you are wondering, he did thankfully recover. But the bullet, fired from a 357 magnum, hit his spinal cord. He will walk with a cane for the rest of his life. An ambulance was just leaving with a student who was hit by a ricochet bullet. The student also recovered.

I remember walking into the cafeteria and seeing the top brass in my school district; superintendent, security director, deputy superintendents, assistant superintendents, all walking in a circle basically saying, "Sixteen year old gunman stood here, assistant principal was shot right here, here's where the ricochet bullet hit the wall and hit the injured student . . ." Essentially, everyone was getting the facts. One thing about crisis intervention is that we need to understand the facts, walk through it so that we'll be able to focus on,

"Okay . . . I understand what happened. Now, what can we do to help everyone with their shock, confusion, anger—the multitude of complex emotions and grief?"

I approached the principal of the school. I didn't know him well. I looked around the cafeteria and wondered, "Where are all the staff members and students who were here only minutes ago? They left behind their lunch trays, their books, their coats, their purses . . . everybody had evacuated. Where have they gone and what are we going to do at this moment to help them?" So I approached the principal with some ideas. I said, "I'd like to talk to you. I have some things, as a psychologist, I think would be important to do right away." He turned to me and said, "Scott, I'm sorry but I don't have time to talk to you right now. Please, write down your ideas and I'll take a look at them." Initially, I thought I was being brushed off, but really my ideas weren't all that earth shattering and anyone reading this narrative would have come up with most, or all of them. The principal actually did every single one of them. They were as basic as gather the faculty, reunite students with their parents, and make a plan for Monday.

The school principal was in great demand! Why was he in such great demand? Well, that sixteen year old gunman, after opening fire in the cafeteria, walked outside and heard the police sirens. When he heard the police in route, he decided he better go back into the school. One person stood between him and reentry. That was the principal. He told me later that he didn't know how he convinced the shooter, but he said, "Son . . . nobody else needs to get hurt. I don't know what's going on in your life but this has to stop. Please, just hand me the gun." So the principal got the shooter to actually surrender. Every reporter in Houston wanted to talk with the heroic principal, who was also besieged by the superintendents, the deputies, the security director, his own assistant principals, counselors, teachers . . . everybody was looking to him to try to figure out what to do next.

No one person, no matter how capable, can possibly be in four or five places at one time. This is why a crisis team . . . members with delegated duties, are essential in schools. Having said this, there was a small nuance I'd like to share that demonstrates how complicated and how important a team response is. The word went out really quickly all over the school district that an assistant principal had been shot at Langham Creek High School. We contacted the injured assistant principal's life partner, who also worked for our school district. She was told what happened by a school counselor in a confidential setting at her school and escorted to the hospital where he was receiving treatment. We also had counselors in the three schools his children attended ready to confidentially tell them what happened to their father, and to support them in every way possible, including transportation to reunite them with their family. We thought of all that. But no one thought of the need to clarify to the families of the *other* three assistant principals at Langham Creek High School that they were *not* shot. "It was not your partner." "It was not your wife." "It was not your mother." "It was not your father." "It was not your husband." These are the details that a crisis team would remember. The word has gone out . . . the critical thing we can do is make sure we give the relevant facts to waiting family members. Inform everyone who needs to know, and most importantly provide clarification for apprehensive, confused loved ones that their family member is safe.

We did conduct a staff meeting after school that fall afternoon. We prayed for our colleague and the student's recovery, answered questions, made plans for the weekend and plans for Monday morning, including a faculty meeting before school. I remember though, that teachers and school staff were all anxious to get home to be with their families. Some of them even commented, "It's a beautiful day outside. I just need to get home." This crisis, by definition, violated any expectations they had about being an educator. It created an incredible sense of disequilibrium, which contributed to just wanting to get home and be with loved ones. I'm sure being with family helped, but it was only the starting point for recovering from seeing their colleague gunned down.

Working in the next few days and weeks with staff and students who were in the cafeteria, they said things like, "Why did I dive underneath the table?" "Why didn't I jump up and knock the gun from his hand?" "Why wasn't I aware of what he was up to?" "Why didn't I somehow prevent this?" There were a lot of guilt-laden "if-only" thoughts. I believe those occur after virtually every school crisis. People just wish fervently that it did not happen, or could have been prevented.

The very next week, there were lots of rumors. Rumors that the shooter was going to make bail. He was coming back to school. It was very important that we respond to these rumors and let the students know he was not going to be released from jail.

His motivation is something that will forever remain unclear. The only statement he ever made was, "The assistant principal stood for discipline. I thought I would do everybody a favor and remove him!" We later learned that he and the assistant principal had only one conversation. At the time of the incident, there was a hair length code for male students, and the administrator had talked to the troubled youth about his hair being too long and needing to be trimmed. That day, the young perpetrator was described as being dressed much like the character in the recently released *Terminator* movie. He told a friend that he had to get something out of his locker. We now know that thing was the 357 magnum brought from his home. He had on a camouflage vest. He put on black gloves. He put on black sunglasses. He went downstairs where the assistant principal was still in the cafeteria and . . . he shot him.

We'll never have all the answers why the assailant did what he did. But it is interesting to look at the judicial process before our states got really tough on juvenile crime and youth mental health needs became recognized as a priority. The injured administrator indicated he felt the school district supported him and his family in every way possible after the tragedy. However, when he went to court and testified at the trial of the sixteen year old, he was made to feel like he had done something wrong. He clearly stated that the assailant receiving a sentence of less than two years and being freed on his 18th birthday, possibly still disturbed and dangerous, was not a sufficient consequence for his act. My colleague, as stated earlier, will walk with a cane for the rest of his life.

PREVENTION AND PREPAREDNESS

Learning and Moving Forward After Two School Shootings

The aftermath of a second tragedy and having two school shootings occur within a period of less than two years in our renowned suburban school district motivated me to do a great deal more. I had already written an acclaimed book, *Suicide Intervention in Schools* (1989) and now, with colleague Gayle Pitcher, we decided to write the book *Crisis Intervention in the Schools* (1992). This book would not only focus on the topic of suicide, but would provide information on other tragic situations, including school shootings, bus accidents, bombings and deaths of students or staff.

In the early 90s, Gayle Pitcher and I started surveying the literature to see what had been written about school crisis. There was not a great deal of information. There was a chapter in a book or two, but not a book that focused solely on school crisis. We were more than a decade away from the National Association of School Psychologists (NASP) creating the PREPaRE model (Brock et al., 2009, 2016) for school crisis prevention and intervention. We realized that Gerald Caplan's (1964) model of crisis intervention was critically important and could be applied to schools. Caplan designated three levels: *primary prevention, secondary intervention,* and *tertiary intervention.* All subsequent chapters and books I have authored or co-authored are based on this model.

First and foremost, what are schools doing to prevent a crisis from occurring? When I was the NASP president in 2000, my focus was accident prevention, suicide prevention, and homicide prevention for all of our children. Preventing these tragedies still remains, many years later, my most passionate mission. I believe our schools are weakest on primary prevention—activities designed to prevent a crisis from occurring such as gun safety and suicide prevention programs. Secondary intervention is what most people think of as crisis intervention—decisions made in the first few hours, activities to help those affected on the day of, and the days after the crisis event. Schools are also weak on what Caplan calls tertiary interventions. What are we doing a month later? What are we doing six months later? What about the anniversary? What about the birthday the deceased would have had? It really raises the question where is the long-term help for those who are severely affected by traumatic events at school.

I approached the superintendent and suggested that having had two school shootings, the district needed to form a crisis task force that would put together policies, training, maybe even evolve to conducting crisis drills. I'll never forget when he looked at me and said, "Scott, I think you need to do it!" Now, there were a couple of things that I think are important. I did accept that responsibility and spent the entire summer working on it with some key people. One of those was the school nurse, who had been at Millsap Elementary when the young man raising the flag was shot. She decided, shortly afterward, that she'd had enough working at the elementary school and transferred to Langham Creek High School. She was there the day that the assistant principal was shot. She was extremely motivated to work on crisis intervention, and it helped that her father-in-law was the school superintendent. Among her contributions, I will forever credit her for making sure our nurses in every school were trained and had standardized emergency kits, including trauma bandages.

I also interviewed a number of our long-time principals and got them engaged in the crisis team planning. When we unveiled our plan for the district, I was able to cite them by name and essentially worked toward a commitment and buy-in from all principals.

The first decision we faced in the district, which is discussed in *Crisis Intervention in the Schools* (1992), was what type of crisis teams we would construct. A district can build *school* crisis teams where everybody works in the same building, having the advantage that everyone knows each other. A second category is called a *district* crisis team, which can be a resource for school teams when a crisis is of such magnitude that they need outside help. Everybody works for the school district, but they play a variety of roles and are drawn from a variety of buildings. District team members usually only see each other for meetings, training, and crises. Small rural districts in this country need what is called a *combined* crisis team, which includes members who work for the school district and others who work in the community. For example, a small district in Texas would be unlikely to have law enforcement or security on staff, and might not even have a full-time counselor or psychologist. Those individuals may have to come from county and community services, and nearby school districts. With this combined model, it is more difficult to get organized and have planning so everyone will be on the same page.

The Cypress Fairbanks district had large elementary schools, middle schools, and high schools so we formed building crisis teams with designated members: law enforcement, campus liaison, medical liaison, counseling liaison, media liaison, and parent liaison. Every one of these liaisons had specific, designated duties that anticipated needs and planned at Caplan's three levels—primary prevention, secondary intervention, and tertiary intervention. The district really embraced this idea.

If you are wondering about the duties of a parent liaison . . . this is not a parent. It is a staff member who communicates with the parents and makes sure they all get the same consistent message. In some of our schools the parent liaison went so far as to train responsible parents living near the school who were available during the day and could come help with an extra set of hands. If we had first graders traumatized by an event and needed comforting from trained adults, this part of the model could be implemented. The building medical liaison, which was the school nurse, identified staff members who knew first aid and CPR and conducted medical emergency drills to assess preparation.

This team approach lead to early writings in my NASP Best Practices chapters. I could go on at length about the duties of these individuals, but the most important thing is that the building crisis team members worked together, frequently met, and even planned crisis drills similar to what are now called table-top drills. It was quite a statement when the superintendent appeared at a school to give the principal a crisis scenario, or required the principal to show documentation of their school crisis team and their plan for drills and exercises! We never used any scary props, never used gunfire or simulated blood. It wasn't difficult to anticipate various school crises, and these exercises, conducted in every building, gave us an opportunity to evaluate our team approach and our readiness to respond at all of Caplan's levels.

RESPONSE TO A FATAL SHOOTING

A Second Shooting at Langham Creek High School

I was feeling good about what we had in place after five years of planning and training on crisis intervention and creating school crisis teams in every single building, when a tragedy occurred. It was again on a Friday in 1992 when we got the call that another shooting had occurred at Langham Creek High School. This time, the incident involved the murder of a teacher that I knew very well. Rita Wenzel was the head of special education at the high school. Unfortunately, she and her husband had been having marital difficulties. One of her colleagues said, "Your husband is outside the school. I know you've been having trouble. I don't think you should go out there." But she did go outside and her husband, who had apparently planned to kidnap her, pulled a gun. Rita ran away from him. We believe she could have actually made it in the school's front door. Every door was locked at that time, except the front door. He started firing at her. She ran right past the front door, made it around to the side of the building, and got her keys in the lock. Sadly, she didn't make it. She was shot dead. Then, he killed himself.

Thankfully, the school had very high windows. There was one group of students who were outside. Girls on the cross country team were taking laps around the building. They saw the murder and suicide. All the other students were inside the building, away from windows, and did not see anything. At that time, I didn't know about the concept and term "circles of vulnerability"—

geographic proximity to the traumatic event, psychosocial proximity meaning closest relationships to the victim, population at risk meaning those with a unique trauma history. So I didn't know the degree of need by "circles of vulnerability," but we were certainly aware that we needed to provide extensive assistance to the girls on that cross country team who witnessed the tragedy.

Thinking back to receiving that call, I remember turning to the school psychologist who was assigned to Langham Creek High School. He turned to me and said, "I'm going home. I'm sorry, because of my personal relationship with Rita, I just can't go to the school and focus on the needs of anyone else." I think a very important point for anyone reading this narrative is that there may be times as a crisis responder that you are simply too close to the situation. Your best response at that time would be not to attempt to lead an intervention to support other people. Instead, sit in the circle where staff is processing, get strength and comfort from your colleagues, from your support system, and it might be a day or two before you are able to focus on the needs of others. It would be helpful for administrators to recognize a responder's need not to lead when emotionally involved to this degree, and to accept that need without judgment or repercussions.

I arrived at the high school and met with the principal. We discussed a number of issues that included not being able to load the bus in the normal location. That's where the two bodies were located. The bodies were actually covered by sheets for several hours. It was also very important to let the students know exactly what had occurred. The principal rehearsed an announcement that was given over the loudspeaker—not generally considered a preferred way of informing. However, to support classroom processing, we were rushing scripts, and notes, and handouts to every single teacher.

The school day ended at the normal time, but our work had barely begun. In the press conference after school, I heavily stressed to parents:

> Langham Creek High School will be open until 10:00 this evening. Counselors and school psychologists will be here. If you have particular concerns about your child, please come to school this evening, bring them and we will assist you and your family.

There was a faculty meeting after school. Then there was a need to contact Rita's family. We were aware that her brother lived in our suburb of Houston. The principal and I went to the home, where we expressed our condolences, asked about funeral plans, discussed with them what could the school do in her memory. It was a very emotional, much needed meeting. I know that her brother and family really appreciated the outreach from the principal and me. We learned that Rita's brother and his family planned to adopt her two small children. Numerous faculty, including the principal and I attended her funeral.

The most affected people at school over the next few days were, of course, her colleagues in special education as she was the team leader of that department. We were very careful to pay attention to their reactions and how they were doing. We had knowledge that some of the members of that team had been dealing with their own previous personal losses and family issues. We also concentrated on the special education students, recognized in the literature as especially vulnerable to traumatic events, and even more affected since Rita had a very close relationship with many of them.

RECOVERY AND REFLECTIONS

One of the things I believe we did best in this situation was a true team response—the result of prior planning and the identification of the various liaison roles. We frequently met with all members of the Langham Creek crisis team to see how they were doing, to anticipate needs and make plans for the future. The principal and every member of that team deserve to be complimented. I just remember reflecting on how different this shooting was. I knew the victim well and liked her. It was shocking that she had been murdered. I found myself wanting to cry, but while at school and working with others, I was able to focus on their needs and control my emotions. But after departing school each day, I cried on the way home. I also had a great need to talk about Rita with my family and colleagues as I reminisced about our interactions and her sense of humor that I always appreciated. It seemed like such a senseless tragedy. Most significantly, this was an actual death that occurred at the school, impacting hundreds of staff and students.

A number of years had passed, seven years to be exact. I was presenting at a statewide conference in Montana on the topic of crisis intervention. After I finished, a gentleman came up to me and introduced himself. Just as he stated his name, I recognized him. He was the brother of Rita. He had moved his family to Montana and had adopted her two young children, now nine and ten years old. It was very good to see him and it was really nice to hear about how her children were doing all these years later. The impact of having somebody that I personally knew shot and killed at school has been very difficult. Even as I am writing this narrative, I have become extremely emotional.

Last year, I was asked to do a keynote address at Emporia State University in Kansas, which is the location of the National Fallen Educator Museum. I toured the memorial, reflecting on the names honored there, and on the fifteen different school shootings that I have personally responded to. I realized one name was missing—a very deserving person—Rita Wenzel, our special education lead at Langham Creek High School. I did not say anything to the curator of the National Fallen Educator Museum because I realize that although shootings at schools still remain, thankfully, incredibly rare, it must be difficult to keep a record and know of every single victim. However, her name deserves to be on that memorial, or at least to be fondly remembered.

CONCLUSION

I have concluded that having these three school shootings occur in my district in a period of just a few years, that my purpose was to respond to them with the best of my abilities. But, more than that, I believe my purpose was to get extensively involved in crisis intervention. I believe I have done that. When I was a finalist for the NASP school psychologist of the year in 1993, I was asked, "What area, what project, what do you believe NASP needs to do that we currently are not?" In my interview and in a written submission, I essentially sketched out that the National Association of School Psychologists needed to become more involved in crisis intervention training and response. In 1995, I was the NASP representative on the U.S. Department of Education Team sent to Oklahoma City after the bombing. I met school leaders from Oklahoma City. One of those leaders, Stephen Crane was on the NASP Executive Committee. He proposed the creation of NASP's National Emergency Assistance Team (NEAT). School psychologists who were experienced in crisis intervention from different regions of the U.S. were selected for the team, which had its first meeting at the 1997 convention. I was the first chairperson of NEAT, serving on the team for 11 years and forming extremely close bonds with the original members. They are special individuals and among my closest friends. I'm very proud to have been involved on the ground floor and am very proud of all of NASP's efforts in crisis intervention, including the creation of PREPaRE. As a result of all these efforts, there are thousands of school psychologists today who are much better prepared to provide support during a school crisis than I was when faced with the Millsap School shooting.

REFERENCES

Brock, S.E., Nickerson, A.B., Reeves, M.A., Jimerson, S.R., Lieberman, R.A., & Feinberg, T.A. (2009). *School crisis prevention and intervention: The PREPaRE model.* Bethesda, MD: National Association of School Psychologists.

Brock, S.E., Nickerson, A.B., Reeves, M.A.L., Conolly, C.N., Jimerson, S.R., Pesce, R.C., & Lazzaro, B.R. (2016). *School crisis prevention and intervention: The PREPaRE model* (2nd ed.). Bethesda, MD: National Association of School Psychologists.

Caplan, G. (1964). *Principles of preventive psychiatry.* New York: Basic Books.

Pitcher, G.D., & Poland, S. (1992). *Crisis intervention in the schools.* New York: Guilford Press.

Poland, S. (1989). *Suicide intervention in schools.* New York: Guilford Press.

16 School Survives a Cycle of Violence

Jeffrey C. Roth[1]

INTRODUCTION

Effects of Repeated Trauma

It seemed the district crisis response team was continually responding to one of our high schools, where students were at risk in the community. Brymer and colleagues (2012) point out that among the key factors influencing how a school responds to a traumatic incident is its history of previous adverse events and experiences with crisis intervention. In this community, gun violence often led to serious injury or death. Sometimes the teenager was making bad choices, engaging in high risk behavior. Other times the youth was innocently caught in the cross fire of violence. If the crisis team felt the burden of three mobilizations to the school within a year, imagine how students from the community felt, under constant threat. To better understand the problem, a brief literature review follows, describing the impact of this cycle of violence on young people.

Children are generally more vulnerable to traumatic events than adults, and childhood psychological trauma can have long term effects on personality development, cognitive and coping abilities (Barenbaum et al., 2004). These high school students had experienced severe emotional trauma throughout their childhood and adolescence.

Pynoos and Nader (1990) addressed the impact of violent loss on urban youth, challenging the myth that young people living in environments with cumulative grief experiences become immune to their psychological impact. This mistaken belief can lead to little or no servicess made available to support recovery. Consequently, the cycle of minimal treatment for emotional trauma "increases the possibility of long-term enduring grief." Crisis teams must have knowledge of the affects of cumulative exposure to violent trauma and the skills to intervene. Jellinek and Okoli (2012) and Jimerson and colleagues (2012) also recognized that students who experienced the sudden, violent death of a relative or close friend were at high risk for psychological trauma (PREPaRE Model: Evaluating Psychological Trauma).

According to Zenere (2009), violence uniquely impacts the urban child's grief.

> "Although theirs is a journey that may be foreign to those not raised in environments scarred by such events, it ... can be contextually understood. Thus, it is critical that the school mental health practitioner be aware of the factors that influence a child's pathway through bereavement."

The impact can be devastating. In fact, the ongoing violence and death in some neighborhoods has been compared with the traumatic experience of children living in a war zone (Bell & Jenkins, 1991).

The condition of repeated traumatic incidents can result in an array of debilitating social, emotional, physical and cognitive symptoms. Parson (1994) has described "urban violence traumatic stress response syndrome", resulting from children's chronic exposure to violence. Zenere (2009) points out that while the child may initially react with shock and disbelief, "with each subsequent loss exposure, affective responses may become increasingly blunted." He describes a litany of research on some of the developmental outcomes, including social withdrawal, aggressiveness, or consuming harmful substances (Lubit et al., 2003). Other outcomes can include dissociation, intrusive thoughts, recurrent dreams, flat affect, or constant motion (Ehrenreich, 2001; ICISF, 2006; Parson, 1994). Children begin burying their emotions, a process that can lead to unhealthy, sometimes violent outcomes (Zenere, 2009).

The preceding studies speak to the potential despair and depression of the inner-city African American student who, year after year, sees friends, relatives, older and younger community members violently die on neighborhood streets. Some of these deaths may be attributed to involvement in the drug trade or gangs or conflicts resolved with weapons, but many involve simply being in the wrong place at the wrong time (Wong et al., 2007). While some schools and communities unite to discourage involvement with drugs and violence, these problems continue to impact young people.

PREPAREDNESS

Preparing for Cultural Competence and the Aftermath of Violence

Athey and Moody-Williams (2003) emphasize the importance of training crisis teams to be aware of the cultural values, traditions, and needs of the diverse school community (PREPaRE Model: Cultural Considerations). They suggest maintaining information about race/ethnicity, languages, history of race relations, history of trauma, and other pertinent data. They point out the benefits of being proactive in identifying community resources, respected leaders, language interpreters, and being prepared to implement solutions to cultural problems arising from a crisis. Crisis teams must also understand how various ethnic groups seek help, cross-cultural outreach strategies, and avoidance of using prejudicial stereotypes or labels. Young (1997) suggested understanding cultural norms about the meaning and expression of emotional pain and grief, and the reactions and rituals associated with death.

Horowitz and colleagues (2005) found that urban African American children and parents clearly preferred seeking social support from family members, rather than professionals who are not trusted. In a study of African American children exposed to chronic violence, Jones (2007) identified kinship support, an Afrocentric perspective, and spirituality as protective factors. Other researchers have connected the church, the religious community, and collaborative religious coping (God and self working together to solve problems) as additional sources of social support and resilience among African Americans (Kim & McKenry, 1998; Molock et al., 2006). Summarizing the research, Brock and colleagues (2016) suggest:

> school-employed mental health professionals seeking to help African American students cope with crises should be particularly aware of the formal kinship support—which may embody values of harmony, inter-connectedness, authenticity, and balance—and of the importance of the Black church.

As crisis responders at the high school, we found that the significance of kinship, expressed in support provided by trusted educators and peers was evident and powerful (PREPaRE Model: Reestablish Social Support Systems).

Schools are hard pressed to prevent violence in the community, but preparing for response to sudden, violent death as part of a comprehensive school safety, prevention, preparedness and response plan is a critical need (Jimerson et al., 2012). An effective program can provide the kind of care and interventions that may help prevent such events in the future (PREPaRE Model: Developing Resilience and School Connectedness: Cultivate Internal/External Resilience).

DESCRIPTION OF RESPONSE

Responding to Homicide in the Community

We had responded several times in one year to a high school where students were exposed to repeated violence in the community. A senior had been killed the previous spring. In the fall, a tenth grade girl was seriously injured in a shooting. Now in spring again, the school endured the shooting death of a popular senior. School counselor Iman Turner first learned of the shooting late in the afternoon from a student he found weeping outside his office.

School administration confirmed the incident she described to him. The injured young man was taken by ambulance to the hospital, where he was placed on life support. His mother returned to Delaware from work in another state. Her son died soon after being taken off life support. The district crisis team was mobilized at the invitation of the school administration. We learned that the victim was the latest casualty in the drug war that plagued his community. Crisis response began the next morning.

The school and district response teams blended for a briefing, prior to a faculty meeting before the start of the school day. An assistant principal and the counselors led the school team. Beginning the briefing, the acting principal reminded the team that in these difficult times, as educators and helping professionals, we needed to take care of each other as well as students and staff. The administrator's reminder contained wisdom that comes from enduring a cycle of senseless and sad experiences. His sensitive leadership was tempered with the knowledge that his staff was beleaguered by repeated trauma. If a team's pressing maintenance needs are not recognized and addressed, the team will soon find it exceedingly difficult to accomplish its task (PREPaRE Model: Caring for the Caregiver). Johnson (1998), commenting on the impact of increased urban violence observed, "Nowhere has this been worse than the inner city, and no one experiences it more closely than the staff of an inner-city school."

Initial Briefing

Some of the main points of discussion and planning at the early morning initial briefing included (PREPaRE Model: Crisis Communication Guidelines):

- Sharing facts as we knew them.
- Discussing what information and supports to provide for teachers and staff at the faculty meeting—information would include a prepared statement to be shared in class with students, anticipated emotional reactions, and the procedure for students needing referral for crisis counseling.
- Discussing what to say, and what not to say to students.
- Preparing for triage and support of students in the library and adjacent rooms, including specific areas for large and small groups, and individual counseling—bottled water, tissues and fresh fruit would be available in the library.
- Brainstorming began generating names of students expected to need support—these included possibly affected students who were in school, out of school, or attending other schools, which needed to be contacted.
- Planning to document names of students receiving interventions, who might also generate names of peers needing support.
- Planning to observe themes and needs of students/staff during the interventions

- Planning to bring the names of counseled students and observed themes to a debriefing tentatively scheduled around lunchtime.
- Periodic debriefings would be ongoing throughout the response, and include the following topics:
  - sharing salient information about the response;
  - sensing needs of students and staff, and planning interventions to meet ongoing needs;
  - planning crisis counseling based upon triage, and assigning responsibility to contact parents/guardians of students most impacted;
  - funeral arrangements and wishes of the family, when available.

- Planning support for the administration, including preparation of letters to the school community by the acting principal, contact with the grieving family, dealing with the media, and arranging for letter copies from district office.

During the briefing, there was discussion of anticipating strong feelings, especially sadness, anger and confusion. The expression of anger related to choices and actions that put the victim at risk would need to be guided in constructive directions. An intervention goal was legitimizing grief for a young man, while encouraging distance from his self-destructive involvement in the drug trade. We also anticipated that the variety of emotional reactions would present teachable moments—for example, the obvious dangers of the drug scene, and having different grief reactions to the same situation.

We talked about response to a similar tragedy less than a year earlier. Applying lessons from that experience would yield mixed results. We anticipated opportunities to form groups in which students could network and support each other (PREPaRE Model: Reestablish Social Support Systems). We also expected opportunities for affected students to seek crisis counseling and support from people they know and trust. We needed a plan to monitor student needs throughout the school's physical plant, including areas difficult to monitor. We considered having at least one trusted interventionist, who lived in the community, available in the large group counseling area of the library. We found that no matter what steps are contemplated, it is simply very difficult, and perhaps impossible to reaffirm perceptions of safety in the face of repeated traumatic violence. In fact, **crisis responders must not assure safety that cannot be assured** (PREPaRE Model: Reaffirm Physical Health and Safety: Providing Accurate Assurances).

Cultural Competence: Boundaries, Barriers and Bridges

As the response unfolded, it reinforced the theorem that, while planning is helpful, events do not always go according to plan. The anguish, the anger, the tension of dozens of African American students from the neighborhood was palpable. They were black and most of the district responders were white. Initially they seemed almost inaccessible to our support. In previous high school responses, it was typical for students to seek comfort from small groups of peers. Crisis counselors could work with those peer groups. The present groups seemed different—their boundaries more guarded, less permeable. Many formed their own tight, private, support system in which grief was clearly evident, but often hidden behind closed mouths and faces devoid of expression. Some of the silence was probably shock, but there were also tears being shed and even a student who seemed to speak incoherently under the influence of drugs or emotional collapse or both. Sometimes in crisis response, the incredible becomes the norm. Several responders noticed a trickle of blood flowing from a crude, homemade bandage around the lower leg of a student. He was sent to the school nurse for treatment of a gunshot wound.

The library, identified for *psychological triage*, was nearly empty. Sporadically, students were directed or trickled in, and there were some instances of individual or small group psychological first aid. In these instances, compassion and empathy from responders knew no racial boundaries or division. However, in this circumstance, large groups of grieving African American students were faced with many counselors whom they did not know. We were strangers to them and most of the strangers were white. I imagine their expectation was that we had little or no understanding of their condition—their sorrow, their despair, their world. We could genuinely care, try to connect, maybe even provide a measure of help by our presence, a word or some brief intervention (PREPaRE Model: Identifying and Responding to Emotionally Overwhelmed Students).

During this difficult time at the high school, it often seemed that skills and caring and cultural competence were simply not enough. I'm not knocking cultural competence. Cultural understanding was definitely helpful, but I'm especially thankful for our African American colleagues, who were sensitive, skilled, and familiar with the students' culture and community. These included the acting principal, two assistant principals, the counselor, and two interventionists, who understood what the students were experiencing and would remain there for them long after the district response team had left the campus. They understood their students on a deeper level than I imagined. I later learned that each of these African American educators had experienced the violent loss of meaningful people in their own lives. Their feeling of kinship with the suffering students was forged in mutual pain and was invaluable (PREPaRE Model: Developing Resilience and School Connectedness: Cultivate Internal/External Resilience).

During the first day of response we learned that a group of 12 to 15 young men and women were huddled close and secure in the small basement office of an interventionist who lived in their community. They chose to be together, protected like a tiny fortress supporting each other, but not accessible to our counseling. We allowed them their safe space and privacy (PREPaRE Model: Reaffirm Physical Health and Safety/Reestablish Social Support Systems). There was a tentative attempt to offer group counseling, but it was difficult in that crowded, rectangular chamber, nearly overflowing as a human barrier to the intrusion of adult intervention. In this situation, what helped most was the presence of adults who had been in these students' lives, whom they knew and trusted—the African American counselor, interventionist, assistant principal, and social worker from our district response team. They took the lead. Other responders tried to be a support system for them.

One of the assistant principals seemed to be "elected" by the situation, by those grieving, by her past experience, to shed tears for them. She had seen too much of this kind of misery and was able to express her grief openly. She periodically spent time with students in the interventionist's room, or invited students who appeared at her office door inside for support. I believe she was helped to keep going by support from two social workers on the district response team. She showed some of us a video she had taken on her cell phone. It was a pep rally where the deceased student performed as M.C. (Master of Ceremonies). He was smiling, handsome, happy, with obvious talent and boundless potential snuffed out by a gunshot. He had made some terrible life choices and now his life ended far too soon.

Eventually, all of the students needing support were encouraged to use the library and adjoining rooms where they could be together or get crisis counseling from those they knew, as well as some of the new faces of the crisis team. The team now had better access to these affected students, at least in order to monitor their emotional status and begin to conduct psychological triage (PREPaRE Model: Secondary Triage: Evaluating Psychological Trauma). This also presented an opportunity to engage in a useful structured PREPaRE intervention, "Student Psychoeducational Group" with many of the students who had entered the library (Brock et al., 2016). During this intervention, some of the students' questions could be answered, rumors could be dispelled, reactions could be discussed and normalized, while performing *secondary triage* to identify maladaptive reactions of students possibly needing further referral. The group was taught stress management strategies. Positive support systems and adaptive coping were reinforced, and students developed stress management plans. The students now had access to art materials, a source of empowerment, and many created and contributed to a memory book for the grieving family (PREPaRE Model: Providing Opportunities to Take Action).

Many of those grieving were serious students, some college bound, and also connected to the neighborhood. They knew who bought drugs, who sold them, who had guns. They understood the criminal subculture. They felt most responsible for not saving their brother. School counselor Iman Turner described some of the critical elements of individually counseling these students. Crisis counseling involved being willing to sit in silence, to listen, to hear their story, to ask questions for clarification and to better understand. Then, to help them reflect on their own questions, to figure out what they needed to do.

In the short term, identified tasks were to grieve, to support the family of the deceased student, and to maintain academic goals. In the long term, they could reject and encourage others to reject the path to drugs and violence, connect with faith-based support, make a commitment to self, family and health, tell family and friends they love them, give everything they have to give to become better people. These themes would help them to mourn another loss, but distance from destructive choices by challenging themselves to strive for fulfilling lives (PREPaRE Model: Providing Opportunities to Take Action).

One of the students seeking counseling in the library stood out because he was different. He was a tall, lanky, white youth, known to be involved in the drug scene. He appeared shaken, perhaps scared, identifying with a young man he probably knew, an unfortunate connection sharing bad choices, now haunted by fearful insights. It could have been him. He entered the library several times during the response, always seeking to talk with Iman Turner. This was clearly an example of a post trauma situation in which the significance of a relationship and trust transcended color and ethnicity.

Triage: Unique Issues of Relative Need

In this crisis response, as in others, there were many nuances of pain and many opportunities for learning. When I recently visited Iman to get his views on the response, he educated me about a unique and challenging aspect of *psychological triage* associated with violent death in the drug counterculture. There were young people in the community who glorified the combat, elevated the violence, viewed the dead teenager as a fallen warrior, a hero. There may have been an element of reverence, but not necessarily grief and mourning. They were certainly stunned and perhaps confused by the incident. They felt connected by the street life, living with the drug selling, using, gun culture. They were shocked and perhaps saddened, but not necessarily in the throes of sorrow like others who grieved. In terms of triage, they were not the priority. Iman would level with them. He respected their feelings, but his priority was to be available for those deeply grieving the loss of a friend, a neighbor, a classmate (PREPaRE Model: Secondary Triage: Evaluating Psychological Trauma).

The counterculture young people appeared almost in awe of the circumstances of the death. They were respectful, but did not display the kind of grief shown by many other students who truly knew the victim as a vibrant, complicated person who had made bad choices. Some of those who grieved had grown up with him. Others had been with him through four years of high school, getting to know and appreciate his hopes, dreams and aspirations. Now they grieved dreams not deferred, but snuffed out of existence.

Iman elaborated on the need to explore the condition of the counterculture youth, whose interest ranged from curiosity to worship, but also possibly grief.

He respectfully explored their feelings with them, sorting out those who truly needed grief counseling. He talked with them honestly about differences in various reactions to the death. And for those caught up in the mystique of drugs and violence, he exercised a profound teachable moment. The drug culture, with its money, materialism and temporary highs, is extremely seductive to poor, hopeless urban youth. However, the drug culture turns out to be fickle, as the temporary high turns into long term addiction. The fast money turns into a target on your back. The truth of the drug culture is concealed. Its promise of affluence is a lie. It is ultimately a killer of hopes. It is a killer of young men and women.

Iman took the opportunity to plant the seed, hoping it would grow. He left the students with a challenge. Reflect on this death. Reflect on your lifestyle. Contemplate change. Think about making choices that are life affirming, rather than self-destructive. "You can come back and talk with me. You don't need to die," he said. Through all of this, Iman himself was grieving and frustrated. Perhaps more of these senseless deaths could be prevented.

RECOVERY

Insights: Inclusion for a Bereaved Community

At the funeral, the deceased student was acknowledged with affection, and remembered with care to distance from and not glorify the actions that precipitated his death (PREPaRE Model: Special Considerations When Memorializing an Incident). The aftermath of the funeral signaled a reasonable return to normal routine at the high school.

After the response, an assistant principal shared some insights and concerns from her perspective. We discussed some of the frustration of trying to embrace grieving students under circumstances in which bad, ultimately fatal choices led to a tragic homicide. We were responding to a situation devastating to a community, but we were sheltering students from facts they already knew. She pointed out that we need to be careful not to restrict information only to those we think may have been impacted. Under difficult circumstances we may try to protect and to shelter, but we need to be honest with the facts and include all students.

The assistant principal talked about being approached by white students after the response. They also felt a connection, the loss of a classmate, regardless of circumstances or culture or color. Perhaps for some, this death evoked feelings associated with another loss of a family member or a friend. They felt "out of it". They felt they were not given the opportunity, invited to grieve. It is understandable that the focus of attention was directed toward the urban African American students, but the response also needed to be more inclusive. It could have reached out to other students troubled by the tragedy (PREPaRE Model: Examine Effectiveness of Crisis Prevention and Intervention).

Valarie Molaison (2003) dispels the myth that, when a student dies, we should inform only those most closely affected.

> When someone dies in the school community, it affects the entire community. True, some will be affected more than others, but tragic or untimely deaths are often met with strong emotional reactions, even in those who did not know the person. Further, it is not possible to discern who will or will not be closely affected by the death. Sometimes people have links with one another outside of school that are not known by school officials.

Saltzman and colleagues (2001) emphasize the need to collect information about students from multiple sources, using multiple methods. They provide a school-based screening and group treatment protocol for adolescents exposed to community violence.

The assistant principal also expressed the belief that, in a crisis, racial barriers between responders and students are not glaring. Students experiencing grief are generally receptive to care, comfort and support that is provided. However, cultural sensitivity is critical. The assistant principal raised an interesting question, "How would I, as a black crisis responder, feel intervening with all Asian American students and families grieving the death of an Asian American student?" The same question could be asked of a white responder with all Asian American students. Could I be helpful? Yes. Would I be at a distinct disadvantage having a gap in knowledge of the Asian experience, culture, way of grieving? Yes. Could I be more helpful if I had knowledge of, and sensitivity to the Asian experience and culture? No doubt, but this is not an easy task, because "Asian" itself is a multicultural term encompassing many nationalities and ethnicities.

In retrospect, I believe that there are many ways to demonstrate cultural competence. It is helpful when the larger culture embraces and values other cultures, while accommodating to their norms. There can be expressions of respect for individuals and their culture. I believe it also helps to be interested and willing to ask, to learn about and understand cultural differences. I will never underestimate the power of genuine empathy, respectfully shared across cultural traditions and values. Whenever possible, we must strive to utilize resources such as school and community leaders, crisis responders, and language interpreters who are a trusted source of comfort (PREPaRE Model: Specific School Crisis Planning Issues: Cultural Considerations).

*   *   *   *

Some years later, I returned to the high school at the end of a school day to interview an administrator. As I entered, one of the counselors, a familiar friendly face, stood in the hallway in front of the door to the counseling suite.

She looked extremely serious, pressing a cell phone to her ear. As I walked past, I thought I heard her say, "killed" into the cell phone. Perhaps after all these years, my ears were attuned to tragic discourse. Though I believe she said it softly, the word seemed to reverberate loudly through the corridor. As I approached the library, school staff were congregating. I gazed through the double doors and inside were some of the members of the district crisis response team—my colleagues, my friends. As the counselor with the cell phone entered the library, she told me a male student had been shot and killed during a robbery attempt in the community.

I declined her invitation to enter the library. I was now retired. I was no longer a team member. It would have been inappropriate to go inside the library. They were a cohesive team with strong leadership. But my feelings were mixed. Relieved I did not have to be involved, yet wanting to be there. Feeling a need, a pull to be with the team. My heart was with the team. Part of me will always be with them.

LESSONS LEARNED AND REFLECTIONS

1. Repeated, violent traumatic incidents in the community can have a cumulative, debilitating effect on school students and staff. *What steps can be taken by a crisis team to help students and staff cope with the terrible impact of repeated violence and death?*

2. There are clear benefits for the crisis team to understand cultural norms such as the expression of emotional pain, reactions to trauma, rituals, and sources of coping and social support among urban African American youth. Perhaps most beneficial is having trusted school and district responders who are African American. *How can a crisis team develop an understanding of cultural norms and the value of diverse team membership, and what are the benefits?*

3. It is expected that violent or deadly outcomes of criminal or high risk behavior will generate confused, contradictory feelings such as both sadness and anger in grieving students. *How do crisis counselors legitimize the expression of grief for a young man killed in drug related violence, while encouraging distance from, and constructive anger toward his destructive choices and activities?*

4. The library set up for triage and interventions in the present scenario was initially ignored by students in favor of the small office of a trusted African American interventionist, where the security of peer support was available. *How can students' need for peer support and security be respected, while encouraging reduced defensive boundaries and access to all available resources?*

5. Triage after urban violence may present unique challenges, such as having to distinguish the different needs of those who truly grieve the death, and youth who are more in awe of a fallen "warrior". *How are very different needs prioritized and addressed in triage and crisis counseling?*

6. There was a false, unstated assumption in the present case, that only the urban African American students needed to grieve and receive support. *Without minimizing the anguish and needs of the grieving African American students, how can all students in the school community be invited to grieve, to recognize the death of a fellow student, and receive crisis intervention if needed?*

The following is an edited letter to parents and guardians after a homicide.

School District/School Letterhead   Date

Parents and Guardians of [School Name] High School Students:

On Friday night, [Date], [School Name] High School lost a member of its family to an act of senseless violence in the city of _____. [Name of Student], a member of the class of _____, was to graduate with his class on June 1. Instead he will be mourned by many and buried in the coming days.

Today our students, faculty and staff observed a moment of silence and grew a little closer in memory of a classmate, friend and student. The faculty and staff along with the district crisis team, Wellness Center staff, counselors, school psychologist and nurse met this morning in advance of our students' arrival. Throughout the day all of our students and staff were offered the opportunity to speak directly with a counselor to openly express their sadness and grief.

I speak from the unfortunate experience of losing a loved one, when I say that death, as a subject of discussion, is extremely difficult with a child. If you as a parent or guardian realize that your child is experiencing grief as a result of this tragic event, I urge you to seek counseling either through referral by your private physician or by contacting one of our own guidance counselors at [School Name] High School, [phone number]. Our counselors have done an outstanding job of listening to our children and providing guidance and support. The mental health and physical well-being of our students is paramount always. We will continue to offer support throughout the week.

Please contact me directly or any member of the administrative staff if you have any concerns or questions. I thank you once again for your continued support for [School Name] High School, its students, faculty and staff.

Sincerely,

Acting Principal

NOTE

1.   Adapted from School Crisis Response: Reflections of a Team Leader (Roth, 2015).

REFERENCES

Athey, J. & Moody-Williams, J. (2003). *Developing cultural competence in disaster mental health programs: Guiding principles and recommendations.* Washington, DC: US Department of Health and Human Services.

Barenbaum, J., Ruchkin, F., & Schwab-Stone, M. (2004). The psychosocial aspects of children exposed to war: Practice and policy initiatives. *Journal of Child Psychology and Psychiatry, 45,* 41–62.

Bell, C., & Jenkins, E. (1991). Traumatic stress and children. *Journal of Health Care for The Poor and Underserved, 2,* 175–188.

Brock, S.E., Nickerson, A.B., Louvar Reeves, M.A., Conolly, C.A., Jimerson, S.R., Persce, R.C., & Lazzaro, B.R. (2016). *School crisis prevention and intervention: The PREPaRE model* (2nd ed.). Bethesda, MD: National Association of School Psychologists.

Brymer, M.J., Pynoos, R.S., Vivrette, R.L., & Taylor, M.A. (2012). Providing school crisis interventions. In S.E. Brock & S.R. Jimerson (Eds.) *Best practices in school crisis prevention and intervention* (pp. 317–336; 2nd ed.). Bethesda, MD: National Association of School Psychologists.

Ehrenreich, J.H. (2001). *Coping with disasters: A guidebook to psychosocial intervention.* Mental Health Workers without Borders.

Horowitz, K., McKay, M., & Marshall, R. (2005). Community violence and urban families: Experiences, effects, and directions for intervention. *American Journal of Orthopsychiatry, 75,* 356–368.

International Critical Incident Stress Foundation (ICISF). (2006). *A murder in the family.* Retrieved June 27, 2009 from: www.thisisawar.com/GriefMurderFamily.htm

Jellinek, M.S. & Okoli, U.D. (2012). When a student dies: Organizing the school's response. *Child and Adolescent Psychiatric Clinics of North America, 21,* 57–67.

Jimerson, S.R., Brown, J.A., & Stewart, K.T. (2012). Sudden and unexpected student death: Preparing for and responding to the unpredictable. In S.E. Brock & S.R. Jimerson (Eds.) *Best practices in school crisis prevention and intervention* (pp. 469–483; 2nd ed.). Bethesda, MD: National Association of School Psychologists.

Johnson, K. (1998). *Trauma in the lives of children: Crisis and stress management techniques for counselors, teachers, and other professionals.* Alameda, CA: Hunter House.

Jones, J.M. (2007). Exposure to chronic community violence: Resilience in African American children. *Journal of Black Psychology, 33,* 125–149.

Kim, H.K., & McKenry, P. (1998). Social networks and support: A comparison of African Americans, Asian Americans, Caucasians, and Hispanics. *Journal of Comparative Family Studies, 29,* 313–334.

Lubit, R., Rovine, D., Defrancisi, L. & Eth, S. (2003). Impact of trauma on children. *Journal of Psychiatric Practice, 9*(2), 128–138.

Molaison, V. (2003). School survival kit: Helping students cope with grief in the school setting. Supporting Kidds. Retrieved from www.supportingkidds.org.

Molock, S.D., Puri, R., Matlin, S., & Barksdale, C. (2006). Relationship between religious coping and suicidal behavior among African American adolescents. *Journal of Black Psychology, 32*, 366–389.

Parson, E.R. (1994). *Inner city children of trauma: Urban violence traumatic stress response syndrome (U-VTS) and therapists' responses.* Retrieved June 28, 2009, from: www.gift fromwithin.org/pdf/parson.pdf

Pynoos, R.S., & Nader, K. (1990). Children's exposure to violence and traumatic death. *Psychiatric Annals, 20*, 334–344.

Saltzman, W.R., Pynoos, R.S., Layne, C.M., Steinberg, A.M., & Aisenberg, E. (2001). Trauma-and grief-focused intervention for adolescents exposed to community violence: Results of a school-based screening and group treatment protocol. *Group Dynamics: Theory, Research and Practice, 5*, 291–303.

Wong, M., Rosemond, M.E., Stein, B.D., Langley, A.K., Katoaka, S.H., & Nadeem, E. (2007). School-based intervention for adolescents exposed to violence. *The Prevention Researcher, 14*(1), 17–20.

Young, M.A. (1997). *The community crisis response team training manual* (2nd ed.). Washington, D.C.: Office for Victims of Crime and the National Organization for Victim Assistance.

Zenere, F.J. (2009). Violent loss and urban children: Understanding the impact on grieving and development. *NASP Communiqué, 38*, (2).

Section 7

School Crisis Response to Child Maltreatment

INTRODUCTION

Definitions

While rarely sensationalized like traumatic events such as school shootings, child maltreatment is a far more prevalent crisis that must be confronted by school personnel. According to the Centers for Disease Control and Prevention (CDC), child maltreatment includes all types of abuse and neglect of a child under age 18 by a parent, relative, or other caregivers (e.g. clergy, coach, teacher) that results in harm, potential to harm, or threat of harm. Acts of commission include physical, sexual, and psychological abuse. Acts of omission include physical, emotional, medical and dental, and educational neglect, inadequate supervision, or exposure to violent environments (CDC, 2015).

Prevalence and Impact

CDC (2016) estimates that one in four children experience some form of child abuse or neglect in their lifetimes. Child Protective Services (CPS) data shows 702,000 victims of child abuse and neglect, with more than 1,500 deaths reported in 2014. It has been estimated that nearly one in ten students suffer some form of verbal or physical sexual misconduct by school personnel during their academic careers (GAO, 2014). An investigation of 2001–2005 educator disciplinary records found that teaching credentials of 2,570 educators had been revoked, denied, surrendered, or sanctioned as a result of sexual misconduct (Irvine & Tanner, 2007).

A study found that the rate of psychiatric diagnoses was 50% in women and 47% in men who had suffered childhood sexual abuse (Martin et al., 2004). Childhood abuse can result in many problems, including anxiety and post-traumatic stress disorders, conduct disorder, problems with adult intimacy, and difficulty with learning, attention, and memory (Canton-Contes & Canton, 2010; Dallam, 2001).

Perpetrators

While commonly believed that most perpetrators are strangers, they are typically someone known and trusted by the victimized child. Sexual abuse of girls is generally by a family member such as a parent, stepparent, sibling, grandparent, uncle, or cousin. Sexual abuse of boys is most often by a person outside the family such as a coach, teacher, neighbor, or babysitter. Perpetrators may be adults, adolescents, or even prepubescent children.

A *Training Guide for Administrators and Educators on Addressing Adult Sexual Misconduct in the School Setting* (2017) provides useful information about policies and procedures to prevent maltreatment, requirements and suggestions about mandated reporting, training in awareness and prevention, the role of social media and technology, and resources to prevent sexual misconduct by school personnel.

The Internet and social media have increasingly become dangerous mediums used by sexual predators to solicit vulnerable young victims (Wolak et al., 2008). Sextortion can happen online when a "sextortionist" threatens to distribute private, sensitive material unless a victim provides images of a sexual nature, sexual favors, or money.

Victims of Sexual Abuse and Factors Mediating Maltreatment Severity

There are risk factors for being sexually abused, but no typical victim profile (Cruise, 2010b). CDC data indicates that the youngest are most vulnerable, with about 27% of reported victims under age three. While girls are more often abused than boys, it is believed that boys are less likely to report abuse. Children who have poor parental supervision and low self-esteem may be at elevated risk for sexual abuse because perpetrators target children responsive to their attention (Cruise, 2010a).

Factors mediating the severity of victim maltreatment:

- Age of the child when abused—younger are generally more vulnerable.
- Type of abuse—the more extreme, physical, or intrusive, the more damage.
- Closeness of the relationship to the perpetrator—family or trusted person means greater sense of betrayal.
- Duration of abuse—chronic abuse over a long period is more disruptive, with the possible exception of short-term trauma involving violence or sadism.
- Emotional climate and reactions of family members—disbelief can increase damage, compounding the betrayal by denying help.
- Mental and emotional health of the child—personality, temperament, and experiential factors such as self-esteem, secure attachments, and resilience are protective and support coping.

Adapted from Webster & Browning, 2002

Schools as a Protective Factor

Schools can have a positive impact on the wellbeing, resiliency, and academic outcomes of abused and neglected children. School experiences can be a protective factor, promoting self-esteem, self-efficacy, opportunities for success, and the development of social, problem solving, and coping skills (Dezen, Gubi, & Ping, 2010; Gonzalez, 2006). School administrators must establish an explicit policy that supports mandated reporting of suspected maltreatment (Cruise, 2010b). School psychologists, counselors, and social workers can raise awareness, educate, and consult with teachers, administrators, staff and parents about the challenges and needs of students traumatized by abuse and neglect. Dezen and colleagues (2010) suggest that school psychologists can help prevent further maltreatment and strengthen the functioning of students affected by abuse and neglect by engaging them in *safety planning*, empowering them with *socialization skills* and social-emotional learning (SEL), teaching them *problem-solving and coping skills*, educating them about abuse and *self protection*, teaching older students about *parenting and life skills*, and *training and supporting teachers* to better understand child maltreatment, recognize warning signs and reporting mandates, and create a safe, supportive environment for their students.

PREVENTION

Helping Children and Families Thrive

The Centers for Disease Control and Prevention (2016) emphasize making environments and relationships safe, stable, and nurturing to help children and families thrive.

Suggestions to help prevent child maltreatment:

- Educate and encourage positive parenting skills to promote healthy child development.
- Support community and school involvement with families to strengthen resources and economic supports.
- Provide quality child care and education early in life.
- Support community and school programs designed to prevent abuse and protect children from potential perpetrators, including on the Internet and social media.
- Provide awareness, prevention, and intervention training on child maltreatment for school personnel, parents, and guardians (GAO, 2014).
- Know the warning signs of maltreatment in order to better recognize when a child is being harmed (Shakeshaft, 2013).
- Be ready to report suspicion of maltreatment to prevent potential for further harm.

- Teach children self-protection skills, about potential abusers, that they have the right to say "no" or "stop", and to tell a trusted adult if they feel at risk, and keep telling trusted adults until they are believed.
- Adult caregivers (e.g. parents, teachers) who have trusting relationships with children should offer ongoing communication about topics such as personal boundaries, appropriate and inappropriate touching, Internet safety, and should remain accessible if children have questions or wish to disclose.

Adapted from Cruise, 2010a; Webster & Browning, 2002

IDENTIFICATION

Warning Signs of Maltreatment

Identifying a child experiencing maltreatment is not obvious, because there is no common pattern of indicators of abuse or neglect. However, there are a variety of physical, behavioral, and psychological indicators that could herald maltreatment. Since these warning signs could indicate other individual or family problems, educators and other adults must be cautious not to immediately assume maltreatment, but to consider symptom history or patterns (Cruise, 2010a). When indicators are sufficient to suspect maltreatment, educators are legally mandated to report their suspicion to designated social service or law enforcement agencies.

Warning signs that may indicate child abuse or neglect:

- Pattern of physical injuries or indicators such as fractures, bruises, burn marks, sexually related medical conditions, sleep disturbances, enuresis or encopresis, constipation, or self-injurious behavior.
- Sudden changes in mood, behavior, or school performance; inappropriate stories or drawings of a sexual nature.
- Difficulty regulating emotions, including change or fluctuation in a child's emotional state such as unusual tearfulness, anger, constricted affect, or outbursts.
- Overly compliant, passive, or withdrawn behavior.
- Depression or loss of interest in pleasurable activities.
- Anxiety which may result in fidgety, distractible, or compulsive behaviors.
- Problems with learning or concentration that are not attributed to specific physical or psychological causes.
- Watchful or hypervigilant behavior, increased irritability, hyperactive and regressive behaviors, seductive behaviors, or flashbacks of a traumatic event.
- Avoidance of intimacy due to feeling vulnerable or loss of control, accompanied by anger, anti-social, oppositional, and other behaviors that promote avoidance.

- Physical or medical problems brought to parents' attention, but not addressed.
- Lack of appropriate adult supervision.
- Comes to school or other activities early, stays late, and does not want to go home.

Adapted from Cruise, 2010a; Cruise, 2010b;
Webster & Browning, 2002

Supporting Disclosure and Reporting Suspected Maltreatment

Some children may find disclosure of maltreatment, and the events that follow more traumatizing than the acts of abuse or neglect (Webster & Browning, 2002). Since children may feel more vulnerable when they open up during a forensic interview, there must be assurance that afterward they will get the protection and treatment they need (Clay, 2014). Rather than false reassurances, school professionals can reduce a child's anxiety by explaining what to expect and what will be done to address the problem (Webster & Browning, 2002).

Suggestions when a child discloses maltreatment:

- Meet the legal mandate to report the maltreatment.
- Listen calmly without overreacting—offer a supportive, nonjudgmental response.
- Reduce victim self-blame, assuring the child that she or he did the right thing to tell what was happening.
- Provide developmentally appropriate information about the reporting process, including the role of the police officer and social worker from protective services.
- Reassure the child that you will continue to be supportive.
- Focus on the experience of maltreatment, assuring the child that she or he is not to blame for the abuse or neglect.
- Reasonably assure that responsible adults will work to protect the child's safety and security, as well as any other children threatened by the perpetrator.

Adapted from Cruise, 2010a; Webster &
Browning, 2002

INTERVENTION

Supporting Maltreated Children

Interventions seek to stop the maltreatment, preclude maladaptive coping and symptom development by helping children express their feelings, lessen their distress, and reinforce adaptive coping strategies (Webster & Browning, 2002).

Suggestions to support maltreated children:

- Provide a safe, healthy recovery environment at home and school, remembering that the involvement of at least one caring, significant adult in the life of the victimized child is essential for positive outcomes.
- Help the child and family find appropriate community resources and treatment.
- Each child may have different responses to abuse or neglect, and may need different interventions from adults to support recovery.
- Be respectful of personal boundaries, especially when there has been physical or sexual abuse—the child's body has been violated so it is imperative to be sensitive to issues of personal space and touching.
- Be prepared to reassure a child manifesting self-blame that the abuse or neglect was not her or his fault.
- Reestablish trust and promote appropriate adult-child relationships— maltreated children may behave in ways that try to reenact the abuse, testing and evoking frustration or anger from caring teachers and other adults.
- Provide clear, consistent limits to help children perceive the environment as safe—an abused or neglected child may feel a lack of control so limits can help the child feel more secure and regain a sense of personal control.
- Enhance self-esteem, promoting positive peer relationships, and supporting academic success and constructive activities.

Adapted from Cruise, 2010a; Webster &
Browning, 2002

Supporting Families of Maltreated Children

Suggestions to support families and caregivers of maltreated children:

- Take your child for a medical exam with a doctor knowledgeable about abused or neglected children—may involve collection of evidence and testing for sexually transmitted diseases and pregnancy.
- Seek counseling with a specialist in child abuse or trauma if needed, for the victimized child, for yourself, or for the family.
- Acknowledge your own feelings about the maltreatment such as guilt, anger, sadness—if you experienced abuse or neglect as a child, you may have emotional reactions that affect the way you respond to and parent your child.
- Balance self-care and childcare—practicing self-care will model how to remain calm and constructively manage emotions or distress for your child.

REFERENCES

A Training Guide for Administrators and Educators on Addressing Adult Sexual Misconduct in the School Setting (2017). Washington, D.C.: Readiness and Emergency Management for Schools (REMS), U.S. Department of Education.

Canton-Contes, D., & Canton, J. (2010). Coping with child sexual abuse among college students and post-traumatic stress disorder: The role of continuity of abuse and relationship with the perpetrator. *Child Abuse and Neglect, 34*(7), 496–501.

Centers for Disease Control and Prevention (CDC). (2015). *Child maltreatment: Definitions.* Retrieved from: www.cdc.gov/violenceprevention/childmaltreatment/definitions.html

Centers for Disease Control and Prevention (CDC). (2016). *Child abuse and neglect prevention.* Atlanta, GA: Author. Retrieved from: www.cdc.gov/violenceprevention/childmaltreatment/index.html

Clay, R.A. (2014). Prosecuting child sexual abuse. *Monitor on Psychology, 45*(8), 50–51.

Cruise, T.K. (2010a). Sexual abuse of children and adolescents. In A.S. Canter, L. Z. Paige, M.D. Roth, I. Romero, & S.A. Carroll (Eds.) *Helping children at home and school III: Handouts for families and educators.* Bethesda, MD: National Association of School Psychologists.

Cruise, T.K. (2010b, September). Identifying and reporting child maltreatment. *Principal Leadership, 11*(1), 12–16.

Dallam, S.J. (2001). The long-term medical consequences of childhood trauma. In K. Franey, R. Geffner, & R. Falconer (Eds.), *The cost of child maltreatment: Who pays? We all do.* (pp. 1–14). San Diego, CA: Family Violence and Sexual Assault Institute.

Dezen, K.A., Gubi, A., & Ping, J. (2010). School psychologists working with children affected by abuse and neglect. *NASP Communique, 38*(7).

Gonzalez, R.A. (2006). Toward the school as sanctuary concept in multicultural urban education: Implications for small high school reform. *Curriculum Inquiry, 36*(3), 273–301.

Government Accountability Office (GAO). (2014). Federal agencies can better support state efforts to prevent and respond to sexual abuse by school personnel. Retrieved from: www.gao.gov/assets/670/660375.pdf

Irvine, M., & Tanner, R. (2007). AP: Sexual misconduct plagues U.S. schools. *The Washington Post.* Retrieved from www.washingtonpost.com/wp-dyn/content/article/2007/10/21/AR2007102100144.html

Martin, G., Bergen, H.A., & Richardson, A.S. (2004). Sexual abuse and suicidality: Gender differences in a large community sample of adolescents. *Child Abuse & Neglect, 28*, 491–503.

Shakeshaft, C. (2013). Know the warning signs of educator sexual misconduct. *Phi Delta Kappan, 94*(5), 8–13.

Webster, L. & Browning, J. (2002). Child maltreatment. In S.E. Brock, P.J. Lazarus, & S.R. Jimerson (Eds.) *Best practices in school crisis prevention and intervention* (pp. 503–530). Bethesda, MD: National Association of School Psychologists.

Wolak, J., Finkelhor, D., Mitchell, K.J., & Ybarra, M. L. (2008). Online "predators" and their victims: Myths, realities, and implications for prevention. *American Psychologist, 63*(2), 111–128.

17 Betrayal of a School Community

Jeffrey C. Roth[1]

INTRODUCTION

The crisis team was mobilized late in the school year by phone tree originating early in the evening from district office. We were told to report the next day to an elementary school for an early morning briefing. An emergency faculty meeting would follow, before the arrival of students. A sixth grade teacher had been charged with multiple rapes of one of her students. The detailed description in the newspaper of the charges against her, was clear and compelling. The *News Journal* reported that the alleged perpetrator acknowledged having sexual intercourse with the student on multiple occasions.

PREVENTION AND PREPAREDNESS

Could this egregious betrayal of trust have been prevented? Painstaking and necessary procedures, including criminal background checks are required in an effort to assure that predators do not find their way into schools. The accused teacher apparently had no previous record of child maltreatment. Schools establish universal programs to create a positive, caring climate. Schools develop comprehensive safety plans for the prevention of various crises, and effective preparation and response to traumatic incidents (Reeves et al., 2010). Schools generally do *not* have response plans for molestation of a student by a teacher. Such an act is so unusual, so beyond comprehension that subtle warning signs of inappropriate teacher behavior might be missed. Still, having a comprehensive, flexible response framework, adaptable for unexpected traumatic incidents proved beneficial in planning the present response (Jimerson et al., 2012).

During the primary grades, many students are presented with programs that discuss *good touch—bad touch*, recognizing when a stranger or a person in your life is not respecting private physical or emotional boundaries, and what to do about such an intrusion. Since these formal efforts to prevent child sexual abuse often involve discussion of body parts and of inappropriate touching, parents may choose to opt out, or educate their children privately (Ortiz & Voutsinas, 2012). Here, there was no such option as sexual terms and behaviors publicized in newspapers violated an entire school community. Because teachers are trusted, warnings rarely identify them as potential predators. The overwhelming majority of teachers have earned a place of respect in society that is beyond reproach. When a rare incident occurs, the act of child maltreatment by a teacher is such an abuse of authority, such a betrayal, that it devastates the community.

Preparing for Crisis Response

A crisis team leader, I lay awake the evening before, and then early on the morning of the response, contemplating the work of the team and the anticipated chaos at school. I expected that our intervention would need to be system-wide to a greater extent than most responses. We had often dealt with the profound grief of a death. However, death does not encompass manipulative malevolence designed by a teacher. She had been charged with multiple counts of rape over a period of time. This incident had severely shaken the sense of safety and security that is the foundation of learning. I anticipated that our intervention would need to address three major areas— the *students*, the *staff* and the *school community*. I did not anticipate the extreme degree to which these groups were impacted. As we learned more about what this person had done, we learned that multiple and varied interventions were needed.

I arrived at the school about 6:20 a.m. and entered the long, rectangular disciplinary room where our team briefing would take place. Approaching a large blackboard at the far end of the room, I wrote three words in large, upper case letters across the top. The words were "STUDENTS", "STAFF", and "COMMUNITY". Under these headings, the team would develop interventions. I brainstormed some of the major tasks under each heading. The affected children would be provided with *psychological triage* and offered crisis counseling in the library. Some smaller rooms adjoining the library could serve as space for small group and individual interventions. A similar process of triage would focus on the teaching staff (PREPaRE Model: Primary Triage: Evaluating Psychological Trauma).

DESCRIPTION OF RESPONSE

I arranged several long tables in a row. Team members and school administrators began entering the room. Bottled water and fresh fruit were brought in. I sat at the head of the table, in front of the words on the blackboard. The principal, dealing with many difficult issues, was in and out of the briefing room. Fortunately, several key members of the school staff were also members of the district crisis team. These members were an assistant to the principal, the school psychologist, and a social worker. The three were highly skilled, professional and caring individuals. They were clearly affected, but able to function effectively and share leadership roles. This made it easier to blend the school and district teams. We were not yet aware of the PREPaRE model (Brock et al., 2009, 2016) and had no formal Incident Command System, but applied concepts such as *division of labor* to address the many required tasks, and *span of control* as leadership arose within small groups addressing a variety of response functions.

Discussing psychological first aid in schools, Brymer and colleagues (2012) emphasize the importance of recognizing and engaging a school's *internal support systems*. The skilled leadership of those who worked within the building and concurrently district level responders, was a huge benefit, particularly when it became evident that *conflict resolution* was necessary. One school counselor was an effective response partner, but another counselor who had a close professional relationship with the accused teacher, was distraught and needed support and a break from crisis related duties.

During this crisis, I took a more authoritative leadership role than I recall in any other response. In the midst of the chaos we would navigate, I sensed that steady, directive leadership was needed. Both crisis responders and school staff needed to feel secure in order to focus on what needed to be done. There was a critical need to maintain open communication among team members, including discussion of information at scheduled debriefings. The sharing of facts, feelings, perceived issues and needs in the school *and* on the team, was necessary for effective planning of interventions. My role included facilitating discussion and planning, and what I sensed was a critical need to provide a safe structure to support the team and do our job. The tension was beyond description. I was constantly monitoring myself and trying to do the right thing for the team. I also monitored the demeanor of the team—the verbal and nonverbal feedback that gave cues when we were on the right track or confused and needing to "back up" and re-evaluate.

Initial Briefing Session

During our initial briefing session, I set the stage asking the administrator and others to share facts about the incident with the team and to gauge the anticipated impact—feelings, issues, needs within the school. The three focus areas on the blackboard provided a useful framework for our discussion and planning. We brainstormed interventions as a group, under each category on the blackboard. We spoke about anticipated themes when working with the students. We needed to hear their stories, understand, empathize, respond to their questions and confusion, but provide reassurance that most adults in school can be trusted (PREPaRE Model: Reaffirm Physical Health and

Perceptions of Safety and Security). We wanted to convey the message that what this teacher did was extremely unusual. We discussed empowering students if appropriate, by providing developmentally suitable information about personal boundaries and protecting themselves from sexual advances by strangers or people they know.

We discussed support for the faculty and prepared for the faculty meeting that now seemed moments away. The principal would take the lead, providing her staff with information about the incident, what to say and not say to students, crisis counseling and support available for students and staff, and tips on how to address the needs of students and parents. Extra substitute teachers were available to relieve faculty if needed. We heard that members of the alleged perpetrator's teaching team were particularly upset.

By now the media was present. The story had gone national. Newspaper reporters and trucks with satellite dishes lined the road in front of the school. Media were directed to the district spokesperson. There would be opportunities to speak with the principal, but media people were prohibited from entering the school building or grounds (PREPaRE Model: Media Relations Protocols/ Collaborating With the Media). The superintendent had joined our briefing and we all proceeded to the faculty meeting, where we dealt with feelings ranging from disbelief and denial to extreme sadness and anger.

*   *   *   *

We did not know the extent of the damage until we began crisis counseling with the students and teachers (PREPaRE Model: Provide Interventions and Respond to Psychological Needs). The sexual acts had occurred in her apartment, often with a second boy observing, but there were many layers of victimization, beyond what anyone imagined. The perpetrator had nearly a full year to victimize the entire school, manipulating her students and creating an unusual, dysfunctional classroom situation.

In her classroom, she apparently had "favorites", including the two boys she had preyed upon off campus, and the girlfriend of the victim. There were reports of inappropriate behavior and dancing in class. She assigned "enforcers" and a "sergeant-at-arms"—students who forced classmates to follow the "rules" and not tell anyone outside the classroom what was happening. For her innocent pupils, class was often a lively, fun place. They had no idea that in a sense, it was an extension of her crime scene. We learned that contrary to school policy, the class was often conducted behind a locked door, with windows covered. Noise sometimes brought complaints, but this fed into the "us" versus "them" mentality encouraged by the perpetrator. We were distressed to learn that she used this "divide and conquer" control tactic not only to bond with her students, but also with her teaching team. The effectiveness of her emotional manipulation became abundantly, disturbingly clear as the crisis response unfolded.

Responding to Students

During crisis counseling, the team learned what had happened in the classroom. Children were in a state of confusion and disbelief. Many were torn by feelings of affection for their teacher and feelings of either denial or disgust at her actions. I recall a distraught youngster saying, "She's the best teacher I ever had." Through tears, he repeated over and over, "She's the best teacher I ever had." What does one say? Usually, we celebrate students holding a teacher in high regard. When this person was referred to as "best teacher", I felt a sense of revulsion. "Your teacher is a pedophile," I thought, but did not say. Not then. I felt outrage toward the perpetrator, but did not show it. Not then.

She had raped his classmate. Didn't he realize? I also felt a mixture of sadness and concern for her influence on this youngster's psyche. Circumstances had shattered his reality—his sense of trust. He must have felt utter confusion seeing someone so highly regarded, charged with a terrible crime. He did not seem to feel any of the anger I felt—only a sense of sorrow and loss. He was saying, "She's the best teacher. . ." in the present tense. How long would it take him to change his estimation, to realize the truth? I listened. I empathized (PREPaRE Model: Individual Crisis Intervention Elements: Establishing Psychological Contact). I recognized that in the classroom, there were ways she seemed like a "best teacher." I tried gently to create some dissonance. It was my need in the moment–his need I felt, in the long run.

> It is very painful . . . it hurts a lot when you find out that someone you like has done some very bad things . . . It takes time and it is hard to accept that someone you think a lot of, is not the person you thought she was . . . She did some things that were very wrong . . . This is a sad time, but after a while you will begin to understand it better and it won't hurt so much . . . There are family and people in school who care about you and will help you begin to feel better.
>
> (PREPaRE Model: Ensure Perceptions of Safety and Security: Providing Crisis Facts and Adaptive Interpretations)

Many parents were contacted personally. Ways to be supportive were suggested. In some cases, further counseling was recommended.

A disturbing theme began to arise during counseling with small groups of students—some from the perpetrator's homeroom, others who had her for one or two subjects. Some students began blaming the victim. Others were silent—whether in tacit agreement that the victim was to blame, or disagreeing, but perhaps afraid to speak in his defense. We began hearing, "It was his fault." "He made her do it." Did some silent students feel their teacher was wrong, but were scared to say it?

We were dealing with a complex, twisted situation. Some students stood in defense of an authority figure they had been taught to respect and obey.

The scenario was further complicated by the victim's status as one of the teacher's "favorites." Students talked freely about the favoritism, the classroom "pecking order", and may have resented the victim's status. He had been receiving treatment in class that went far beyond the boundaries of "teacher's pet." The perpetrator's manipulation had elevated her target's status in relation to her, while alienating and isolating him from his peers.

There was division between students who recognized that what their teacher did was not acceptable, and students who defended her and denied culpability. Still others were silent, fearful, confused. We began intervening with homogeneous groups having similar beliefs, and with individuals, in order to encourage open discussion of feelings with less concern about peers judging one's perceptions. For those who blamed the victim, the response needed to be direct. The adult is *always* the responsible party and must always do the right thing regardless of the child's actions. A teacher, like a parent, has a special responsibility to keep her students safe. Was it right for a teacher to have "favorites"? Was it right for a teacher to use her position of authority to take advantage of a student and cause harm? The answer to both questions was a resounding, unequivocal, "NO".

Reestablishing Trust

Working with the students, it was imperative to speak honestly about this teacher doing something very wrong, while reassuring them about the trustworthiness of responsible adults, including teachers. Their world was torn by fear and mistrust. How is trust reestablished, along with a healthy dose of caution and good judgment? Listening. Information. Discussion. Education. We clearly informed students that what this person did was not the normal way that teachers and adults act. We assured them that the vast majority of teachers and adults in positions of responsibility for children's wellbeing do the right thing. We discussed what a young person can do "if something doesn't seem right" relative to personal boundaries. "If you are suspicious about someone's words, actions or intentions, talk with your parents, guardians, and trustworthy adults." Communicate with those you feel you can trust. Ask questions. Find answers. Find that you can trust again (PREPaRE Model: Ensure Perceptions of Safety and Security).

Enlisting Systems of Support to Keep Students Safe

One of the many difficult aspects of crisis counseling was dealing with sexual terms brought up by students. For instance, the newspaper had used the term "oral sex". There was a disparity of knowledge among the children. For many, terms like "sexual intercourse" and "rape" were foreign and confusing. They continued to be victimized, robbed of their innocence by circumstances set in motion by the sick and selfish acts of an adult entrusted with their care. As crisis counselors, this was one of the most frustrating conditions. We felt that the children shouldn't be learning about these terms in this way. . .not in this context. It just wasn't fair. We wanted to keep the students safe, but in this instance, we felt powerless to prevent exposure to more trauma (PREP<u>a</u>RE Model: Minimize Exposure to Trauma).

Never before had the provision of psychological first aid felt like a descent into hell. We found some time during the day to vent, but mostly to debrief, share concerns, talk about the themes, and structure interventions for the students. There was little time to deal with our own feelings. With the school counselor taking the lead, we engaged in *secondary triage*, generating names of students who were at risk for traumatic stress reactions (PREP<u>a</u>RE Model: Secondary Triage: Evaluating Psychological Trauma). Crisis team members shared the responsibility of contacting these students' parents or guardians by phone to consult with and educate them about reactions to trauma, discuss ways to support their children, and when to consider referral for therapeutic treatment.

Throughout the response, it was critical that families and especially parents/guardians of students in the perpetrator's classroom be kept informed. Letters from the principal were sent home with students at the end of the day, explaining the situation and the response, providing tips on how parents could be supportive, inviting them to contact the school with concerns, and informing about referral for counseling if needed. Finally, carefully planned parent meetings would be held at the school. Communication at every stage enlisted cooperation, helped calm fears, created a spirit of collaboration and increased goodwill (PREP<u>a</u>RE Model: Reestablish Social Support Systems/ Caregiver Training).

Early on the first day, while planning for afternoon interventions, the team discussed how to get students home safely through the gauntlet of media waiting outside. We called in district administrators to help secure the perimeter and create a buffer between the media and the students. It was relatively easy to cover the boarding of school buses. The routes taken by those walking home from school and the various bus stops along the way were more difficult to shield. The large number of administrators provided from the district office was invaluable. They were placed at strategic locations along the walking routes and at key bus stops. This support was appreciated by the crisis team and helped keep the students safe (PREP<u>a</u>RE Model: Prevent Trauma Exposure: Keep Students Safe).

Responding to Staff

The Second Day: Layers of Victimization

Another layer of victimization added to the staff's pain and anger over what had happened. The perpetrator was white. The targeted victim and the observer were African American. The reality of race in this crime was recognized and discussed by some, ignored by others. As a crisis team working with a multi-racial staff, there was limited discussion of the racial issue, but some African American staff members were especially upset, expressed anger, and questioned the way this abuse happened. There was potential for more divisiveness among a staff already divided, as we would learn, by the accused teacher's manipulation. Administration and staff needed an opportunity to express and resolve feelings around the issue of race, but I'm not certain whether it happened during or after the immediate crisis, except for individual discussion and private complaint.

Conflict Resolution for Teaching Teams

Several members of the sixth grade team were extremely affected, but generally insisted on staying with their classrooms. Crisis counselors spoke with them individually, seeking emotional stabilization so they could care for their students (PREPaRE Model: Ensure Perceptions of Safety and Security: The Effect of Adult Reactions and Behaviors). We supported other staff, including members of the fifth grade team, who were affected in a different way. We were surprised to learn that there was animosity between some members of the fifth and sixth grade teams, with hard feelings traced back to manipulation by the perpetrator. The "us" versus "them" control strategy had been used with fellow teachers as well as students. Then, we learned that it was continuing. Out on bail, the accused was sending emails and phone messages to the sixth grade team. I suggested we plan an intervention to work first with the fifth grade team as a group, and then the sixth grade team as a group. Each team was comprised of five or six teachers.

We began planning an intervention to address the needs of a sixth grade team experiencing grief, disbelief, confusion, and even protective feelings toward a colleague who could not have done these acts. She was the center of a tight social network that included her team, but excluded others. It appeared that her priority was to have fun and if necessary, bend school rules doing it. Stretching the rules and relating to students more like peers made it more difficult for other teachers to set limits. This seemed especially true for some members of the fifth grade team, who were feeling anger, resentment, and concern for students they had previously taught. They also needed a special intervention. Before the incident, communication between teams appeared closed, except for complaints directed at the sixth grade team, who defensively banded together. News of the rapes created an even more divisive situation. A wound festered among staff that would make healing from the emotional trauma more difficult.

We planned separate group interventions, including an element of *conflict resolution* for the members of each teaching team, beginning with the fifth grade team (PREPaRE Model: Psychoeducation). Members of the school and district crisis teams, the school psychologist and the assistant to the principal, would facilitate separate, closed meetings for each team. The group crisis counseling sessions would be closed to responders not on the school staff. The co-facilitators, school psychologist Phyllis Tallos and assistant to the principal Wilma Robinson agreed that the intervention was needed and readily accepted their roles. The principal and school administration helped plan the logistics, and the sessions were scheduled in the crisis team planning room for the following day.

The entire crisis team discussed how each of the sessions would be structured. The district's assistant superintendent sat in on the planning, listening attentively, silently. The teams would first need an opportunity to tell their stories—how they learned about the alleged rapes, what they thought, how they felt. They would be encouraged to describe the most difficult aspects of the event from their perspective. They would have an opportunity to hear thoughts and feelings they had in common—to see that their reactions were typical, given very abnormal circumstances. They would then be encouraged to talk about ways to cope with the intense stress that weighed upon them. The facilitators would summarize what they had learned and then provide some important information. They would be reminded about stress reduction strategies, networks of support, and individual counseling availability. Finally, each team, fifth and sixth grade in their separate meetings, would be encouraged to understand the perspective, the struggle of the other. This was a crucial part of the intervention that we hoped would begin to heal the schism between the two teams. The sixth grade team was struggling with grief, loss, and protective feelings that accompanied a sense of disbelief. They were likely feeling isolated, defensive, and vulnerable. The fifth grade team was feeling understandable anger, resentment, powerlessness, exclusion and perhaps a degree of guilt for some of these feelings. The teams would likely share similar feelings—especially an overwhelming concern for the students.

Teacher Team Sessions

On the third day of the response, time passed slowly. As the time to begin the first teacher team session approached, it appeared that the assistant superintendent intended to enter and observe. She confirmed that this was her intention. I requested that she not sit in on the teacher team sessions, explaining, "They need to meet as staff, facilitated by colleagues, without being observed. They need privacy and confidentiality to speak freely, without censoring their feelings, or fearing what is said will be repeated outside the group." She got on her cell phone. I assumed she was calling the superintendent for direction. What he said would determine whether or not she entered the sessions. To his credit and with my gratitude, he told her to stay out.

It was very difficult not to participate in these staff counseling sessions. I checked in with Phyllis and Wilma after each session. They thought the sessions were helpful, the teams receptive—even appreciative. I felt incredible appreciation for Phyllis and Wilma, honored to work with them. I trusted them.

Responding to Community

The crisis team set about focusing on the third major part of the system— the community (PREPaRE Model: Caregiver Training). With school and district administration, we planned an evening community meeting in the school library, where we had triaged and counseled many students. It felt like the room rippled with tension and discomfort. There were parents, guardians, and others who were afraid, angry, anxious about their children, and full of unanswered questions. Many of these questions could never be answered because we either did not know the answers or needed to respect confidentiality. There were legitimate questions about some of the decisions made or not made during the school year. The superintendent, school principal, and director of human resources spoke to the crowd. I don't recall much of what they said. I do recall that the father of a student tore into the superintendent with a verbal barrage of accusations and blame. Often during response to traumatic incidents there is free floating anger. It may be expressed indiscriminately and directed off target, often toward an authority figure. Someone must be blamed. The perpetrator wasn't invited. The superintendent was vulnerable, bearing the brunt of misdirected anger. The man's verbal attack continued for an inordinate amount of time. The superintendent responded, the barrage continued, became repetitive. I remember thinking, "This does not feel like a safe atmosphere. Enough." From the back of the room I loudly, but calmly stated in the direction of the man who attacked, "Sir, you've had a chance to make your points. Let's give other people a chance to speak." Brief silence. "Sorry," he said, and the meeting moved on.

RECOVERY

For many at the community meeting, including myself, emotional recovery began when Phyllis, the school psychologist, acknowledged and expressed the extreme anger so many of us felt and then molded it in a constructive direction. "I'm sure many of us would like to put our hands around the throat of that person and strangle her, but that wouldn't be lawful or the right thing to do," she said. The immediate reaction varied from uninhibited laughter to stunned silence. The collective response was tension release. It was a cathartic statement, to say the least. "What we can do is take care of our children and each other. . . ." Phyllis made suggestions and pointed to a prepared handout that had been circulated. One of the parents, a psychiatrist from DuPont Hospital for Children, had volunteered to speak about children's needs in this unusual situation. She generously offered counseling for children in need, through the hospital.

* * * *

Our blended crisis response team—both school and district, met for a fairly extensive debriefing session shortly after our work at the school was finished (PREPaRE Model: Caring for the Caregiver). We processed the response, shared feelings, vented, talked about stress management. It helped, but was not enough. About a year later we met again to devote another debriefing session to exorcising this event. Again, it helped, but was not enough. I believe many of us have come to terms with the fact that our attempts to care for each other were helpful, but they could never be enough.

Having descended into the depths of hell to address a school community's despair, I was left with the question, "How do we ascend from the depths of hell?" After some reflection, I believe we ascend on the shoulders of our determined colleagues–fellow responders and educators whose acts of kindness show children that there is still goodness in an imperfect world and that a measure of trust can be restored.

LESSONS LEARNED AND REFLECTIONS

1. Depending on the severity of a traumatic event, be prepared to plan for system-wide interventions. *Who are the system-wide stakeholders and what types of interventions might be needed?*
2. The leadership style to facilitate a response should be appropriate for the needs of the situation. While shared leadership and input are always relevant, styles may range from more democratic, consensus-driven to more directive decision making. Styles may require flexibility, varying with different tasks during the same response. *What are some leadership styles and decision-making processes, and when would they be most appropriate?*
3. Given the rare example of teacher misconduct and manipulation, the crisis counselor may work with students having high regard for a person who has committed reprehensible acts. *How are students' feelings respected while the counselor begins to create some dissonance in their view of the teacher, eventually enabling a more reality based and critical viewpoint?*
4. In situations where there is a perpetrator and a target, prepare for the possibility that the victim will be blamed, and the need to clearly ascribe responsibility to the perpetrator. *What are some strategies for working with groups and individuals on a developmentally appropriate understanding of the dynamics of abuse?*
5. During crisis response involving sudden death and instances of physical or emotional abuse or manipulation, there may be a need to conduct

conflict resolution. *What are some ways crisis teams can develop conflict resolution skills and recognize when their use is needed?*

6. Keeping the school staff and community informed and seeking appropriate feedback can make the difference between people cooperating with or obstructing constructive efforts. *What are ways to keep staff and caregivers informed and supportive of recovery efforts?*

Here's a template for a letter sent to parents.

School Letterhead   Date

Dear [School] Parents,

Earlier today, a teacher at [School name] Elementary School, Ms. _____, was arrested and charged with multiple counts of sexual misconduct with a student.

    The District's first concern is to provide additional support for all students at [School]. A District team of counselors, psychologists, and social workers are assigned to the school to support the [School name] counselors and staff as they address the needs of all students. If you have any questions or concerns about your child's reaction to this situation, please contact _____, Assistant to the Principal, at [phone number].

    In the best interest of students, the District has building substitute, Mrs. _____assigned to Ms. _____'s classroom. The District will recruit a high quality teacher as soon as possible for the remainder of the year.

    We know that you will join us in our concern and support for the students and staff of [School name].

    Please direct media inquiries that you may receive to _____, District Information Officer at [phone number].

Sincerely,

_____, Superintendent

_____, Principal

SAMPLE AGENDA: MEETING WITH PARENTS, GUARDIANS, COMMUNITY

- Introductions
- Overview of Agenda
- Description of the event (and its resolution, if appropriate)
- Describe steps which have been taken at school in regard to safety, security and emotional support for students, families and staff.
- Describe common reactions to this type of event.
- Describe ways adults can help their children cope with the situation.
- Describe indicators of need for referral for more intensive support.
- Describe services available at school and in the community and how to access these services.
- Question and answer period
- Close with a statement of hope and compliments for the many strengths evident in the community and how they are pulling together to help one another.
- Recognize anger or other feelings of the audience, but redirect the focus of the meeting to helping students return to school and learning.

NOTE

1. Adapted from *School Crisis Response: Reflections of a Team Leader* (Roth, 2015).

REFERENCES

Brock, S.E., Nickerson, A.B., Reeves, M.A., Jimerson, S.R., Lieberman, R.A., & Feinberg, T.A. (2009). *School crisis prevention and intervention: The PREPaRE model.* Bethesda, MD: National Association of School Psychologists.

Brock, S.E., Nickerson, A.B., Louvar Reeves, M.A., Conolly, C.A., Jimerson, S.R., Persce, R.C., & Lazzaro, B.R. (2016). *School crisis prevention and intervention: The PREPaRE model* (2nd ed.). Bethesda, MD: National Association of School Psychologists.

Brymer, M.J., Pynoos, R.S., Vivrette, R.L., & Taylor, M.A. (2012). Providing school crisis interventions. In S.E. Brock & S.R. Jimerson (Eds.) *Best practices in school crisis prevention and intervention* (pp. 317–336; 2nd ed.). Bethesda, MD: National Association of School Psychologists.

Jimerson, S.R., Brown, J.A., & Stewart, K.T. (2012). Sudden and unexpected student death: Preparing for and responding to the unpredictable. In S.E. Brock & S.R. Jimerson (Eds.) *Best practices in school crisis prevention and intervention* (pp. 469–483; 2nd ed.). Bethesda, MD: National Association of School Psychologists.

Ortiz, S.O., & Voutsinas, M. (2012). Cultural considerations in crisis intervention. In S. E. Brock & S.R. Jimerson (Eds.) *Best practices in school crisis prevention and intervention* (pp. 337–357; 2nd ed.). Bethesda, MD: National Association of School Psychologists.

Reeves, M., Kanan, L., & Plog, A. (2010). *Comprehensive planning for safe learning environments: A school professional's guide to integrating physical and psychological safety— Prevention through recovery.* New York, NY: Routledge.

Section 8

School Crisis Response

Current and Future Directions

18 Current and Future Directions in School Crisis Response

Terri A. Erbacher and Melissa A. Reeves

Narratives throughout this text have addressed multiple types of crisis events, steps taken to respond, team member roles, and the impact of these critical incidents on students, staff, faculty, administration, and crisis team members. Crisis team leaders shared their perspectives on successes, challenges and pitfalls, noting the importance of self-reflection during and after a response. The previous chapters provide many lessons learned and highlight the importance of ongoing skill development and implementation of supports to help all impacted by a crisis. As we look to future directions in crisis response, prevention and trauma mitigation strategies including the role of technology and social media, school-wide supports such as social-emotional learning, creating trauma-informed environments and effective reunification procedures, and ensuring care for caregivers, are areas that need further attention.

THE ROLE OF TECHNOLOGY AND SOCIAL MEDIA

Impact of Social Media

With social media at the fingertips of nearly every middle and high school age child (and many at the elementary level), students often hear of crisis events before staff or administrators, leaving little time for school crisis teams to strategize a coordinated response. Robinson and colleagues (2015) suggest that social media can quickly provide outreach to a large number of individuals, the ability to monitor and intervene with those indicating suicidal ideation or expressing threats, and provides a platform to share experiences, including grief. Multiple narratives throughout this text address varied roles social media played in crisis response, including the forum it provided to quickly share important information as well as challenges of rumor control, cyberbullying, exposure to graphic images, and the role of social media in luring/grooming for abuse, trafficking, and sexual assault. In the introduction to section 4, the role of a social media manager to cope with the challenges associated with social media usage is addressed.

Social Media Management

Establishing a technology team (Brock et al., 2016) at the district and/or school level that includes the social media manager, instructional technology department, public information officer, and members of the school's crisis team can be critical to managing social media challenges. Technology teams can be prepared prior to a crisis situation to help ensure that confidential procedures for reporting concerns are established and operational (e.g. text line, hotline). Ongoing meetings will help the team stay on the forefront of trends, plan for how crisis information will be disseminated, and create a social media presence to build a network base (Brock et al., 2016). Engaging student members can be helpful as they are often technologically savvy, know the popular sites among peers, and often see concerning posts or remarks before adults. Prevention and preparedness also includes educating the public on warning signs, sharing resources, promoting help-seeking behaviors, and decreasing stigma attached to events such as suicide, drug overdose, and the shame bestowed upon many abuse survivors.

Technology teams can manage social media networks to share crisis-related information, control rumors, and identify and monitor students at risk. Technology teams can also help limit exposure by encouraging students not to make videos or share graphic details online, and by encouraging digital breaks from social media posts and photos or memories of the deceased, particularly for youth exposed to trauma. After a suicide death, technology teams are key in promoting *safe messaging* by working with local press not to sensationalize the suicide and to educate the public on mental illness, warning signs, resources, and to provide messages of hope and resiliency (Erbacher, Singer & Poland, 2015). More information on social media and crisis preparedness can be found via the PREPaRE model (Brock et al., 2016), including suggestions for evaluating the use of technology and social media after each crisis.

Contagion

One of the biggest challenges of social media usage is cyberbullying and abusive or harmful postings that can lead to contagion behaviors. Trends via social media change quickly, though cyberbullying remains rampant with 72% of 12–18 year olds reporting they experienced cyberbullying within the last school year (NCES, 2013). The introduction to section 7 discusses the increase in sextortion, whereupon social media predators use blackmail and threaten victims to send incriminating photos or videos (Saul, 2016). Other current trends include livestream violence or suicide and pro-suicide or self-harm games and sites (Instagram: #iwanttodie, #secret_society123, #bluewhale). The ability to livestream video has led to streaming violence and suicide, subsequently exposing viewers to graphic content. This creates widespread potential for contagion (Getz, 2017). Similarly, exposure and contagion are concerns with the Netflix series, 13 Reasons Why (#13RW), as experts worry that this show portrays suicide as a solution to problems, blames survivors for a suicide, feeds revenge fantasies, shows the graphic death of a teen named Hannah, and glorifies her suicide, making Hannah appear larger in death than in life. Concern has also been raised that the adults in the show (counselors, parents) do not take appropriate action to thwart Hannah's suicide, which may leave teens feeling hopeless that anyone can help them (NASP, 2017).

Online Communities: Exposure and Support

Social media can offer an online forum of connection for youth, with implications for both positive and negative outcomes. Vulnerable students may connect on sites that relate to depression, suicide, or self-harm (#deb, #sue, #selfharmmm). Therefore, social media sites must have lifesaving measures whereupon users can report posts, comments, or photos that indicate potential suicidality, allowing site administrators to follow up and offer help and resources (Erbacher et al., 2015). Social media sites have the tough job finding a balance between banning abusive content (harmful language can be reported by users or identified by algorithms) and users feeling they are being censored (Hinduja, 2017). Social media sites are increasingly flagging certain phrases to alert users to potentially graphic images and working to quickly identify and remove violent videos to limit exposure. Livestream videos that show a person committing an act of self-harm or threatening to harm others continue to be a significant challenge. They clearly present crisis exposure for viewers, but provide the opportunity to intervene and get the person help. In addition, social media sites must consider that cultural perspectives regarding what words and actions constitute inappropriate content, abuse or hate speech might vary (Hinduja, 2017).

One of the positive uses of social media is the ability for students to post condolences and memorials, which can bring personal healing, create a sense of belonging, and unite the community in the grieving process. Sofka and colleagues (2012) found four main reasons users turn to social media after losing a loved one: connection (staying connected by viewing posts of happy memories); communication (messages on a deceased person's social media page); commemoration (web pages become personal memorials); and continuation of relationship (maintaining the bond). While expressing grief online can help, it remains important for grief counselors to dialogue about online behavior with the bereaved by asking questions about how they characterize their virtual relationship with the deceased (Sofka, Gilbert, & Cupit, 2012).

Apps

Technology teams and school mental health professionals can stay abreast of new, beneficial apps for students to build resiliency, mitigate risk, and promote wellness. For example, The Great Kindness Challenge is a resiliency-building app for ages 4–18 that promotes random acts of kindness to reach a school-wide goal. Mindfulness apps such as Calm, Breathe2Relax, or Aura, send daily reminders to improve meditation habits for building wellness. There are a multitude of anti-bullying apps such as STOPit or Bully Tag, which enable students to anonymously report bullying by sending messages, pictures, or video to trusted adults in their school. Finally, apps that can be used by youth at risk for suicide prevention include Safety Net or My3, both of which provide customizable safety plans to identify individual warning signs and safety strategies and connect with both personal contacts and national resources. Other suicide prevention apps include A Friend Asks (Jason Foundation) and Ask and Prevent Suicide (Mental Health America of Texas), both of which aim at students who might be able to help a peer. As social media, apps, and technology change rapidly, both schools and parents are encouraged to stay on top of what is trending. While apps can be helpful, it remains critically important for youth to tell an adult should they be concerned about themselves or a peer.

Helping Parents Navigate Technology and Social Media

Many parents do not feel they have the technological capability to keep up with their child's online behavior or to protect their children online (Hinduja & Patchin, 2014; Wong, 2011), with Spanish speaking parents feeling the least confident (FOSI, 2014). While phones have parental controls available and parents are interested in learning more about monitoring tools, just 53% report using them and 31% report using other controls such as parental GPS, (FOSI, 2014). While parents may trust their pediatricians or parenting websites for information on protecting their child online, they put the greatest trust in their child's school (FOSI, 2014). However, schools must consider the "digital divide" when considering technology, realizing that some families cannot use certain technologies due to disability or economics (Sofka et al., 2012), or may not have the education or time to learn these skills. Thus, technology teams can post information on what is trending to help inform parents. Schools can also encourage parents to use social media to open lines of communication with their children, building trust and giving them a safe place to disclose feelings and concerns. Parents are encouraged to create their own social media accounts to monitor what their children and their friends are posting, to have their children's accounts and password information, to discuss sexting and teach their children how to protect themselves from predators, and to educate their children on inappropriate content and the importance of being discreet

(Erbacher et al., 2015). Finally, parents are encouraged to contact their school-based mental health professionals or administrators should they have concerns.

SCHOOL-WIDE SUPPORTS

SEL Programming

Resiliency is a protective factor against risk of harm (Brock et al., 2016). Resilient individuals cope better with adversity, thereby lessening the potential for traumatic impact. It is critical for schools to place more emphasis on prevention programming that includes a strong focus on building resiliency. Internal resiliency factors are characteristics within an individual that help them overcome challenging situations (e.g. self-esteem, emotional regulation, good problem solving) and external resiliency factors are characteristics that foster positive relationships and support systems (e.g. school connectedness, engagement in activities, family support). One of the most effective and well-researched approaches (CASEL, 2013, 2015) to enhance resiliency is the implementation of social-emotional learning (SEL), which has been called the "missing piece" of America's education system (Gayl, 2017). Extensive research demonstrates that SEL approaches that teach skills such as self-awareness, self-management, social awareness, relationship skills, and responsible decision-making, lead to long-term academic and career success, and positive school outcomes such as improved classroom behavior, increased achievement scores, higher graduation rates, and increased ability to handle stress (Osher et al., 2016).

The Every Student Succeeds Act (ESSA) includes language that encourages schools to provide staff development and programming that enhances school safety, ensures educational equity, and includes social-emotional programming. While all 50 states have social-emotional learning standards for pre-k, only three have social-emotional learning standards that span all grade levels: Illinois, West Virginia, and Kansas (Blad, 2016). Eight other states (California, Georgia, Massachusetts, Minnesota, Nevada, Pennsylvania, Tennessee, and Washington) recently applied to work with CASEL (Collaborative for Academic, Social, and Emotional Learning; www.casel.org/) to create and implement social-emotional learning in their schools.

Research also shows SEL can help prevent suicide (Stern & Divecha, 2017). SEL teaches youth strategies to manage feelings—particularly negative emotions, without being overwhelmed by them. It also teaches students to bring feelings into balance to overcome tough emotional periods like coping with crisis situations. SEL helps foster social supports and relationships, powerful variables for emotional recovery from traumatic events (Brock et al., 2016). Further, SEL reduces bullying and harassment by fostering a sense of trust, belonging, and connectedness among students, which helps to reduce suicide risk (Stern & Divecha, 2017).

Trauma-Informed Schools

Toxic stress is strong, frequent, prolonged activation of stress mechanisms that can negatively impact physical and psychological wellbeing. This often includes exposure to chronic trauma situations such as physical or emotional abuse, chronic neglect, caregiver substance abuse or mental illness, exposure to ongoing violence, and/or the accumulated burdens of family economic hardship (Harvard, 2017). Toxic stress and chronic trauma impact many students as nearly 35 million children in the United States are living with emotional and psychological trauma (National Center for Health Statistics, 2012). The landmark Adverse Childhood Experiences (ACE) study demonstrated a clear correlation between toxic stress and negative physical and psychological outcomes (CDC & Kaiser Permanente, 1998). The documentary film, *Resiliency—The Biology of Stress & the Science of Hope* (www.youtube. com/watch?v=We2BqmjHN0k), highlights this science and helps further exemplify the importance of schools adopting a trauma-informed/trauma-sensitive approach.

Research demonstrates that schools adopting a universal trauma-informed/trauma-sensitive approach have increased positive academic and behavioral outcomes. Teachers report feeling better prepared to meet the needs of their students (Cole et al., 2005; Cole et al., 2013; NASP, 2015a; Rossen & Hull, 2012; SAMHSA, 2014; http://traumasensitiveschools.org/trauma-and-learning/the-problem-impact/; https://traumaawareschools.org/traumaIn Schools, https://traumasensitiveschools.org/). A trauma-informed approach involves school staff and related service providers being trained on how to recognize and respond to those who have been impacted by traumatic stress. There is a shared understanding among all staff that trauma effects learning, behavior, and school-based relationships. Students' struggles with behavior and emotional regulation are often due to trauma responses to perceived threats. Instead of seeing behavior as rooted in opposition or defiance and the child being asked, "What did you do?", the child is asked, "Tell me what happened?" Trauma-sensitive school educators also make the switch from asking "What can I do to fix this child?" to "What can we do as a community to support all children to help them feel safe and participate fully in our school community?" The school embraces teamwork and staff share responsibility for all students.

Trauma-sensitive schools take a holistic approach to foster growth in four key domains: a) relationships with teachers and peers; b) ability to self-regulate behaviors, emotions, and attention; c) success in academic and non-academic areas; and d) physical and emotional health and wellbeing (SAMHSA, 2014). These domains maximize students' opportunities to overcome adversity in order to succeed at school. In addition, the school connects students to the school community and provides multiple opportunities to practice newly developing skills while fostering a culture of acceptance and tolerance. All students are welcomed and taught to respect the needs of others. Individual support services are also offered to help students effectively problem solve. Students are provided with clear expectations and communication strategies to guide them through stressful situations. Leadership and staff anticipate and adapt to the ever-changing needs of students and take the time to learn about changes and challenges in the local community. Strong parent engagement and communication is also a valued focus (Cole et al., 2005; Cole et al., 2013; NASP, 2015a; Rossen & Hull, 2012; SAMHSA, 2014).

Research is clear that schools adopting a trauma-informed/trauma-sensitive approach are better able to meet the academic and social-emotional needs of their students, and teachers feel more empowered to positively address the needs of students exposed to toxic stress. However, while a review by Overstreet and Chafouleas (2016) found positive preliminary evidence, trauma-focused professional development training in educational environments has yet to be fully evaluated and more outcome measures are needed.

Reunification Planning

Reunification is linked to all phases of crisis preparedness (USDOE, 2013), yet it is one of the most ignored and underdeveloped aspects of school emergency planning. The stakes are high, as a poorly developed reunification plan can potentially become a source of chaos after a traumatic event. While expedient reunification reduces the potential for lasting trauma to both children and their parents (National Commission on Children and Disasters, 2010), poor reunification increases threat perceptions and traumatic impact, and negatively effects academics, behavior, and social-emotional stability (Brock et al., 2016; Gay & Reeves, 2017; Reeves et al., 2011; Vernberg & Vogel, 1993; Welko, 2013). Children who become separated from their parents or guardians are at much higher risk for secondary trauma effects (Brandenburg, Watkins, Brandenburg, & Schieche, 2007; National Commission on Children and Disasters, 2010). Risk also increases with the duration and distance of separation, younger developmental age, and presence of developmental delays (National Commission on Children and Disasters, 2010; Chung et al., 2012).

Because effective reunification is the first step in the recovery process, there should be a stronger focus on reunification planning and the successful execution of reunification protocols in crisis situations. In addition, school-based mental health professionals should provide high quality crisis interventions and mental health supports during and after reunification. According to the PREPaRE model, a variety of crisis interventions can be delivered at a reunification site. These include reestablishing social supports, disseminating information regarding how to help students and adults cope, and providing caregiver training to help adults learn youth support strategies (Brock et al., 2016; Gay & Reeves, 2017). The I Love U Guys and Safe and Sound Schools Foundations have created reunification resources (I Love U

Guys, 2011, http://iloveuguys.org/srm; www.safeandsoundschools.org). Safe and Sound Schools offers workshops on reunification and integrating lessons learned from real life school tragedies (e.g. Sandy Hook) to help districts develop improved reunification protocols and prepare for successful execution. Schools can better attend to the development and research of efficient and effective reunification to inform future practice.

CARE FOR THE CAREGIVER

Another area needing additional research and focus is caring for the caregiver. Providing supports to those impacted by a crisis is physically and emotionally demanding (Brymer et al., 2006; Brymer et al., 2012). School-employed mental health professionals exposed to traumatic events and working with survivors, can experience negative reactions (Bolnik & Brock, 2005; Tosone et al., 2011). The advent of social media makes it more difficult for crisis team members to take breaks, thereby increasing risk for burnout and vicarious traumatization. Vicarious distress in caregivers may include *cognitive symptoms* such as loss of objectivity or an inability to stop thinking about crisis events; *physical symptoms* such as sleep disturbance or somatic complaints; *affective symptoms* such as excessive worry, suicidal thoughts or irritability/anger; and/or *behavioral/social symptoms* such as alcohol/substance use, withdrawal from friends, or an inability to return to normal job responsibilities (NASP, 2015b). Professionals working in schools may be personally impacted while having the challenge of providing crisis interventions concurrent with other demanding job responsibilities and maintaining existing caseloads.

Prevention is emphasized as studies have found that consistent, active participation in wellness activities helps decrease vulnerability to vicarious traumatization (Bober et al., 2006; Bride, 2007; Bride et al., 2004; Bride et al., 2007; Hunter & Schofield, 2006). Administrators and crisis team leaders can support prevention by offering stress management, meditation classes, crisis response training, sufficient crisis staff, regular supervision, peer supervision, case conferencing, and referrals to employee assistance programs (EAPs) when needed (Erbacher et al., 2015). Suggestions to mitigate secondary traumatization during crisis situations include knowing your limits and setting boundaries on crisis-related work, maintaining normal eating and sleeping routines, recognizing that crisis reactions are normal, and processing the events with trusted colleagues (NASP, 2015b). Administrators can help by limiting shifts of responders, rotating responders from high impact to low stress assignments, and by monitoring those who are vulnerable, including those who have experienced prior trauma (Brock et al., 2016). Narratives throughout this text noted situations in which it was helpful to provide crisis team members with food and water, group debriefings and one on one check-ins, opportunities for breaks, and providing administrators and responders permission to care for themselves. In addition to providing individual and group support for caregivers, implementing a trauma-informed/trauma-sensitive approach also provides self-care instruction for educators.

Brock and colleagues (2016) suggest creating self-care plans that include finding a balance between work and home, setting boundaries, accepting things one cannot change, identifying social supports, and seeking ways to "gain recognition for, and take joy in, the achievement in one's work." As school-based mental health professionals advocate for crisis training and self-care, it is imperative to model these skills and find strategies that work. Renewal may come from spending time in nature, finding ways to laugh, time with children or pets, exercising, engaging in relaxation by meditation and mindfulness, ensuring adequate sleep, or engaging in creative arts such as writing, drawing, or music.

It is important to recognize that self-care is a "fundamental necessity" and ethical principles of self-care should be an ongoing process rather than an afterthought (Figley, 2002). By ensuring their own wellness first, crisis team members are better able to help others. Due to the necessity of self-care, in addition to self-care programs schools can provide, more research needs to be done on the impact of crisis response work on school-employed professionals, and how to mitigate negative consequences.

RESEARCH

Conducting research on school crisis intervention and recovery is challenging due to many methodological, practical, and ethical challenges (Brock & Jimerson, 2012; Nickerson et al., 2012). There are ethical and practical limitations to experimentally control and manipulate variables associated with trauma and violence (Nickerson & Gurdineer, 2012). Measuring post-crisis outcomes requires collection of baseline data before a crisis. While academic data is available pre-crisis, universal mental health screenings are not routinely conducted in many schools. Thus baseline data on mental health is often non-existent. In addition, school administrators and district lawyers are hesitant to have research conducted on students and staff after a crisis due to sensitivity to crisis exposure and potential pending lawsuits. Lastly, opportunities for grant funding often require that the intervention will demonstrate a direct correlation to increasing academic achievement. Complex variables involved in learning and challenges measuring long-term follow-up and recovery (Nickerson & Gurdineer, 2012) make the direct correlation of school crisis interventions to academic achievement difficult to prove. It is also difficult to predict where and when a crisis will occur. With strict requirements for conducting human subject research and required approval from a research review board, by the time the study is approved the window of opportunity to capture data is missed.

There is research to support the effectiveness of crisis prevention and intervention. Program evaluation research has been conducted on the

PREPaRE Crisis Prevention and Intervention Curriculum (Brock et al., 2011; Brock et al., 2016; Nickerson et al., 2014; www.nasponline.org/prepare/training-outcome-eval.aspx), demonstrating an increase in knowledge and confidence in applying skills learned in the workshops. Lazzaro (2013) found that, for individuals completing the PREPaRE Workshop 2 (Crisis Intervention and Recovery), a combination of acquiring the workshop content and years of experience in the field resulted in greater retention and application of knowledge and self-reported confidence in providing interventions. Thus, training, opportunities to apply acquired skills, learning from narratives such as those in this book, and real-life experience implementing the PREPaRE model (www.nasponline.org/professional-development/prepare-training-curriculum/prepare-in-practice) can all contribute to enhanced application of crisis intervention skills. While this study is a good beginning, further research is needed to examine the extent to which additional professional training leads to meaningful systemic changes in crisis prevention and preparedness, collaboration, and improved recovery outcomes for those impacted by crises.

SUMMARY

Future directions need to focus on the challenges in managing social media, increased implementation of school-wide programming such as SEL and trauma-informed schools, effective reunification protocols, and self-care interventions to support staff doing this incredibly hard work. While empirical research has challenges, narratives are one of the ways literature describes our experiences, informs our work, and promotes advances in the application of school crisis response training (Brock et al., 2016; Roth, 2015), demonstrating the importance of this book and continued advocacy. Trauma exposure does not have to define survivors when we can help survivors define the trauma in a way that helps them toward the path of recovery.

REFERENCES

Blad, E. (August 1, 2016). Social-emotional learning: States collaborate to craft standards and policies. *Education Week*. Retrieved from http://blogs.edweek.org/edweek/rulesforengagement/2016/08/social-emotional_learning_states_collaborate_to_craft_standards_policies.html

Bober, T., Regehr, C., & Zhou, Y. R. (2006). Development of the coping strategies inventory for trauma counselors. *Journal of Loss and Trauma, 11*, 71–83. doi: 10.1080/15325020500358225

Bolnik, L., & Brock, S. E. (2005). The self-reported effects of crisis intervention work on school psychologists. *The California School Psychologist, 10*, 117–124. Retrieved January 23, 2015, from www.caspsurveys.org/NEW/pdfs/JRNLv_117.pdf

Brandenburg, M. A., Watkins, S. M., Brandenburg, K. L., & Schieche, C. (2007). Operation child-ID: Reunifying children with their legal guardians after Hurricane Katrina. *Disasters, 31*(3), 277–287. doi:10.1111/j.1467-7717.2007.01009.

Bride, B. E. (2007). Prevalence of secondary traumatic stress among social workers. *Social Work, 52,* 63–70. doi: 10.1093/sw/52.1.63

Bride, B. E., Radey, M., & Figley, C. R. (2007). Measuring compassion fatigue. *Journal of Clinical Social Work, 35,* 155–163. doi: 10.1007/s10615-007-0091-7

Bride, B. E., Robinson, M. M., Yegidis, B., & Figley, C. R. (2004). Development and validation of the Secondary Traumatic Stress Scale. *Research on Social Work Practice, 14,* 27–35. doi: 10.1177/1049731503254106

Brock, S. E., & Jimerson, S. R. (2012). Reflections on prevention, intervention, and future directions. In S. E. Brock & S. R. Jimerson (Eds.), *Best practices in crisis prevention and intervention in the schools* (2nd ed.; pp. 731–733). Bethesda, MD: National Association of School Psychologists.

Brock, S. E., Nickerson, A. B., Reeves, M. A., & Savage, T. A., & Woitaszewski, S. A. (2011). Development, evaluation, and future directions of the PREPaRE School Crisis Prevention and Intervention Training Curriculum. *Journal of School Violence, 10,* 34–52. doi: 10.1080/15388220.2010.519268.

Brock, S.E., Nickerson, A.B., Reeves, M.A., Jimerson, S. R., Conolly-Wilson, C, Pesce, R. & Lazzaro, B. (2016). *School crisis prevention & intervention: The PREPaRE model* (2nd ed.). Bethesda, MD: National Association of School Psychologists.

Brymer, M., Jacobs, A., Layne, C., Pynoos, R., Ruzek, J., Steinberg, A., et al. (2006). *Psychological first aid: Field operations guide.* Rockville, MD: National Child Traumatic Stress Network and National Center for PTSD. Retrieved May 5, 2014, from www.nctsn.org/nccts/nav.do?pid=typ_terr_resources_pfa

Brymer M., Taylor M., Escudero P., Jacobs A., Kronenberg M., Macy R., & Vogel J. (2012). *Psychological first aid for schools: Field operations guide* (2nd ed.). Los Angeles, CA: National Child Traumatic Stress Network. Retrieved from www.nctsn.org/content/psychological-first-aid-schoolspfa

CASEL (2013). CASEL Guide: Effective Social and Emotional Learning Programs Preschool and Elementary School Edition. Retrieved from www.casel.org/preschool-and-elementary-edition-casel-guide/

CASEL (2015). CASEL Guide: Effective Social and Emotional Learning Programs— Middle and High School Edition. Retrieved from www.casel.org/middle-and-high-school-edition-casel-guide/

Center for Disease Control and Prevention (CDC) and Kaiser Permanente (1998). Adverse Childhood Experiences (ACE) Study. Retrieved from www.cdc.gov/violenceprevention/acestudy/index.html

Chung, S., Mario Christoudias, C., Darrell, T., Ziniel, S. I., & Kalish, L. A. (2012). A novel image-based tool to reunite children with their families after disasters. *Academic Emergency Medicine,* 19(11), 1227–1234. doi:10.1111/acem.12013; 10.1111/acem.12013

Cole, S.F. Greenwald O'Brien, J., Gadd, M.G., Ristuccia, J., Wallace, D.L., Gregory, M. (2005). Helping Traumatized Students Learn: Supportive school environments for children traumatized by family violence. Massachusetts Advocates for Children: Boston, MA. Retrieved from https://traumasensitiveschools.org/wp-content/uploads/2013/06/Helping-Traumatized-Children-Learn.pdf

Cole, S.F., Eisner, A., Gregory, M., Ristuccia, J. (2013). *Helping Traumatized Students Learn: Creating and Advocating for Trauma-Sensitive Schools.* Massachusetts Advocates for Children: Boston, MA. Retrieved from http://traumainformedcareproject.org/resources/HTCL-Vol-2-Creating-and-Advocating-for-TSS.pdf

Erbacher, T.A., Singer, J.B., & Poland, S. (2015). *Suicide in schools: A practitioner's guide to multi-level prevention, assessment, intervention and postvention*. New York, NY: Routledge.

Every Student Succeeds Act of 2015, Pub. L. 114–95. Retrieved from www.congress.gov/bill/114th-congress/senate-bill/1177/text

Family Online Safety Institute (FOSI). (2014). Parenting in the digital age: How parents weigh the potential benefits and harms of their children's technology use. Research Brief. Retrieved from www.fosi.org/policy-research/parenting-digital-age

Figley, C. R. (2002). *Treating compassion fatigue*. New York, NY: Brunner-Routledge.

Gay, M. & Reeves, M. (2017). After the crisis: Seamless recovery and reunification procedures. Workshop presented at multiple locations.

Gayl, C. (2017). A brief from the Collaborative for Academic, Social, and Emotional Learning. How state planning for the Every Student Succeeds Act (ESSA) can promote student academic, social, and emotional learning: An examination of five key strategies. Retrieved from www.casel.org/wp-content/uploads/2017/04/ESSA-and-SEL-Five-Strategies-April-2017-041717.pdf

Getz, L. (2017). Livestreamed Suicide on Social Media—The Trauma of Viewership. *Social Work Today*, 17 (3),14.

Harvard University (2017). Center on the Developing Child. http://developingchild.harvard.edu/

Hinduja, S. (2017, July 7). Harmful Speech Online: At the Intersection of Algorithms and Human Behavior. Retrieved from https://cyberbullying.org/harmful-speech-online-intersection-algorithms-human-behavior

Hinduja, S. & Patchin, J. (2014). Cyberbyllying: Identification, prevention & response. Retrieved from http://cyberbullying.org/Cyberbullying-Identification-Prevention-Response.pdf

Hunter, S. V. & Schofield, M. J. (2006). How counselors cope with traumatized clients: Personal, professional and organizational strategies. *Journal for the Advancement of Counseling*, 28, 121–138. doi: 10.1007/s10447-005-9003-0

I Love U Guys (2011). Standard Reunification Protocol: A Practical Method to Unite Students with Parents After an Evacuation or Crisis Based on the Adams 12, Five Star School District Practices. Bailey, CO: Author. http://iloveuguys.org/srm/Standard%20Reunification%20Method.pdf

Lazzaro, B.R. (2013). A survey study of PREPaRE workshop participants' application of knowledge, confidence levels, and utilization of school crisis response and recovery training curriculum (Doctoral dissertation). Retrieved from http://ecommons.luc.edu/luc_diss/674

National Association of School Psychologists (NASP). (2015a). Creating trauma-sensitive schools: Supportive policies and practices for learning [Research summary]. Bethesda, MD: Author.

National Association of School Psychologists (NASP). (2015b). Care for caregivers: Brief facts and tips. Retrieved from www.nasponline.org/resources-and-publications/resources/school-safety-and-crisis/care-for-caregivers.

National Association of School Psychologists (NASP). (2017). 13 Reasons Why Netflix series: Considerations for educators. Retrieved from www.nasponline.org/resources-and-publications/resources/school-safety-and-crisis/preventing-youth-suicide/13-reasons-why-netflix-series-considerations-for-educators

National Center for Education Statistics (NCES). (2013). Student reports of bullying and cyber-bullying: Results from the 2011 school crime supplement to the national crime victimization survey. Retrieved from https://nces.ed.gov/pubs2013/2013329.pdf

National Center for Health Statistics at the Centers for Disease Control (2012). National Survey of Children's Health. Interactive data website: http://childhealth data.org/learn/NSCH#

National Commission on Children and Disasters (2010). The 2010 report to the President and Congress. (No. 10-M037). Rockville, MD: Agency for Healthcare Research and Quality. Retrieved August 1, 2017 from http://archive.ahrq.gov/prep/nccdreport/nccdreport.pdf

Nickerson, A. B. & Gurdineer, E.E. (2012). Research needs for crisis prevention. In S. E. Brock & S. R. Jimerson (Eds.), *Best practices in crisis prevention and intervention in the schools* (2nd ed.; pp. 683–699). Bethesda, MD: National Association of School Psychologists.

Nickerson, A. B., Pagliocca, P. M., & Palladino, S. (2012). Research and evaluation needs for crisis intervention. In S. E. Brock & S. R. Jimerson (Eds.), *Best practices in crisis prevention and intervention in the schools* (2nd ed.; pp. 701–730). Bethesda, MD: National Association of School Psychologists.

Nickerson, A. B., Serwacki, M. L., Brock, S. E., Savage, T. A., Woitaszewski, S. A., & Reeves, M. A. (2014). Program evaluation of the PREPaRE school crisis prevention and intervention training curriculum. *Psychology in the Schools*, 51, 466–479. doi: 10.1002/pits.21757

Osher, D., Kidron, Y., Brackett, M., Dymnicki, Jones, S. & Weissberg, R.P. (2016). Advancing the science and practice of social and emotional learning: Looking back and moving forward. *Review of Research in Education*, 1–38. doi: 10.3102/0091732X 16673595

Overstreet, S. & Chafouleas, S.M. (2016). Trauma-Informed Schools: Introduction to the Special Issue. *School Mental Health*, 8:1–6. doi 10.1007/s12310-016-9184-1.

Reeves, M., Nickerson, A., Conolly-Wilson, C., Susan, M., Lazzaro, Jimerson, S., Pesce, R. (2011). PREPaRE: Crisis Prevention and Preparedness (2nd ed.)— Comprehensive School Safety Planning. National Association of School Psychologists, Bethesda, MD. Primary author.

Robinson, J., Cox, G., Bailey, E., Hetrick, S., Rodrigues, M., Fisher, S., & Herrman, H. (2015). Social media and suicide prevention: a systematic review. *Early Intervention in Psychiatry*. doi:10.1111/eip.12229.

Rossen, E. & Hull, R. (2012). *Supporting and educating traumatized students: A guide for school-based professionals*. New York: Oxford University Press.

Roth, J. C. (2015). School crisis response: Reflections of a team leader. Wilmington, DE: Hickory Run Press.

Saul, J. (2016, December 1). Online 'sextortion' is on the rise, Newsweek. Retrieved from www.newsweek.com/2016/12/09/sextortion-social-media-hacking-blackmail-527201.html

Sofka, C. J., Gilbert, K. R., & Cupit, I. N. (Eds.). (2012). *Dying, death, and grief in an on-line universe: For counselors and educators*. New York, NY: Springer Publishing Company.

Stern, R. & Divecha, S. (2017). Emotional Intelligence Education Has a Role in Suicide Prevention. Retrieved from https://journal.thriveglobal.com/how-emotional-intelligence-plays-a-role-in-suicide-prevention-8e81fb0ce204

Substance Abuse and Mental Health Services Administration (SAMHSA). (2014). SAMHSA's Concept of Trauma and Guidance for a Trauma-Informed Approach. HHS Publication No.—SMA# 14-4884. Rockville, MD: Author. Retrieved from https://store.samhsa.gov/product/SAMHSA-s-Concept-of-Trauma-and-Guidance-for-a-Trauma-Informed-Approach/SMA14-4884

Tosone, C., McTighe, J. P., Bauwens, J., & Naturale, A. (2011). Shared traumatic stress and the long-term impact of 9/11 on Manhattan clinicians. *Journal of Traumatic Stress, 24,* 546–552. doi: 10.1002/jts.20686

U.S. Department of Education, Office of Elementary and Secondary Education, Office of Safe and Healthy Students (USDOE). (2013). Guide for developing high-quality school emergency operations plans. Author: Washington, DC. Retrieved from http://rems.ed.gov/docs/REMS_K-12_Guide_508.pdf

Vernberg, E.M. & Vogel, J.M. (1993). Interventions with children after disasters. *Journal of Clinical Child Psychology, 22,* 485–498.

Welko, A. (2013). Bringing Families Together: Parent-Student Reunification Procedures in School Crisis Planning. Retrieved from http://corescholar.libraries.wright.edu/mph/145

Wong, Y. C. (2011). Cyber-parenting: Internet benefits, risks, and parenting issues. *Journal of Technology in Human Services, 28*(4), 252–273. doi: 10.1080/15228835.2011. 562629.

Index